Barns of Old Mission Peninsula and Their Stories

by Evelyn Johnson

Barns of Old Mission Peninsula and Their Stories
by Evelyn Johnson

Copyright © 2007 by Evelyn Johnson
    Cover photo courtesy of Eileen Brys of Brys Estate Vineyard and Winery
    Photographs by Carl and Evelyn Johnson
    Page design and layout by Susan Harring
    Cover design by Tom White

All rights reserved. No part of this book may be reproduced or transmitted in any form or by any means, electronic or mechanical, including photocopying, recording, or by any information storage retrieval system, without written permission from the publisher.

Revised edition published by:

    Book Marketing Solutions, LLC
    10300 E. Leelanau Court
    Traverse City, Michigan 49684
        orders@BookMarketingSolutions.com
        www.BookMarketingSolutions.com

        Printed in the United States of America

Johnson, Evelyn.
    Barns of Old Mission Peninsula and their stories / by Evelyn
Johnson. -- Traverse City, Mich. : Book Marketing Solutions, 2007.
      p. ; cm.

      ISBN-13: 978-0-9790834-1-9
      ISBN-10: 0-9790834-1-9
      Includes bibliographical references and index.

      1. Barns--Michigan--Pictorial works. 2. Old Mission Peninsula
(Mich.)--History. 3. Grand Traverse County (Mich.)--History.
4. Vernacular architecture--Michigan. I. Title.

NA8230 .J64 2007
728/.9220977464   0702

First edition published by Eladybug Publications of Traverse City, Michigan
with assistance from Michigan State University Museum, 2006.

      Please remember to respect property rights and only view barns from the road.

      This book is available at:
      www.ReadingUp.com

TABLE OF CONTENTS

PREFACE—George McManus, Jr. iv

ACKNOWLEDGEMENTS v

INTRODUCTION—Julie Avery, Curator of Rural Life and Culture, MSU Museum vi

PERSPECTIVE—By the Author, Evelyn Johnson vii

A BRIEF HISTORY OF OLD MISSION PENINSULA ix

MAP OF OLD MISSION PENINSULA x

BARNS AND THEIR STORIES
    Chapter One   Barns of 1850-1900 1
    Chapter Two   Barns of The Turn of the Twentieth Century 25
    Chapter Three   1900-1920 Barns 55
    Chapter Four   1920-1994 Barns 91
    Chapter Five   Phantom Barns 119
    Chapter Six   Centennial Barns 149
    Chapter Seven   Restored Barns 169
    Chapter Eight   Kroupa Barns on Kroupa Road 189

FARMING AND COMMUNITY THROUGH THE YEARS
    Farming from 1840's onward—in Words and Pictures 196
    Peninsula Fruit Receiving and Processing Plants 202
    Life of a Peninsula Farm Wife—Mary Johnson 204
    First Fire Truck 205
    Growing Up on Old Mission Peninsula—George McManus, Jr. 206
    Old Mission School Essays—Fourth Grade Students:
        Chris Cooke and Samantha Knudsen 208
    Old Mission Peninsula Leader in Farmland Preservation—Glen Chown 209

BIBLIOGRAPHY 210

INDEX 211

"Build a barn first. It will pay for the house later," was the axiom of the early settlers. The barn was the money generator on the farm. It was the man's workplace…morning and evening and sometimes in between. The younger generation also learned to work there,

> How to milk the kicking cow
>> Or how to herd the wayward sow.
>
> How the working horse to groom
>> How the stalls to scrape and groom.
>
> How to aim to feed the cat
>> And end the day with 'that is that.'

By Senator George McManus, Jr.

ACKNOWLEDGEMENTS

Every barn has a story to tell. Most of the information and the stories in this book are based on oral history, that is, recollections and memories of the people who live on a farm on the Old Mission Peninsula with a barn or have lived on one in the past. While I was not able to get complete histories on any one barn/farm, I do have wonderful bits and pieces of life and living on the Peninsula. There are more barns on Old Mission Peninsula than you will see in this book, but I have tried my best to get most of them.

I have talked to nearly two hundred people, and I am deeply indebted to each and every one who gave me their time and memories, and sometimes loaned me precious pictures of their barns. If some facts are skewed, names inaccurate, or other errors made, know that this is largely an oral, or spoken, history. It was supported by any written historical data I could find. Also, keep in mind, memories and interpretations vary. My goal and interest is to preserve the wonderful heritage of the barns of this unique place in Michigan called Old Mission Peninsula.

Among those who have my heartfelt gratitude and thanks are:

Julie Avery, MSU Museum assistant curator for her support, expert knowledge, friendship, and guidance. She has been my mentor on barns. I probably would not have completed the project without her encouragement. Thanks, Julie! Also, to the MSU Museum for their assistance.

Carl, my husband, who patiently entered all the pictures in the computer, (photographing many of them himself). He also spent hours and hours on the computer working on details and unsnarling my attempts to make the computer work. Also, Carl gave me a great deal of insight on barns and farming, having been a farm boy himself. And, for just putting up with me during this long Barn Book project, I love you Carl.

Many warm thoughts and thanks to Cal and Verla Jamieson, Jack Solomonson, Bob DeVol, and George McManus Jr. for the hours they spent driving around, telling me stories and giving me facts about the Peninsula and barns. Also, to Steve Stier who toured the Peninsula with me helping me understand barn construction.

Also, I would like to thank:

Jack and Vi Solomonson for loaning me pictures of old barns that were included in the book.

Ann Swaney, NMC Librarian, for the tedious job of indexing the book. I am most grateful for all the work and time she has devoted to make sure the names were accurate and pages indexed properly. She is a gracious lady who was most patient with me. Also to Ann's co-worker, Lynn Freeland, who helped set up the references and indexes, thank you.

Gil Uithol and Sally Akerley, Peninsula Township Assessors, who helped me greatly with dates and facts about Old Mission barns.

Mary Johnson, Carol Lewis, and Fred Dohm Sr. for their enthusiastic support and help.

Richard and Mary Ellen Rabine for pictures from their Blessings and Barns collection, soon to be a calendar of Barns of Old Mission.

Blossom Jerue, who at 93 has a fantastic memory of the past on the Peninsula. We became good friends and I enjoy her company immensely.

I wish to thank Greenstone Farm Credit Services and Brys Estate and Winery for their financial support.

A special thanks to Julianne Meyer for the factual information I retrieved from her book, *Reflections of Yesterday, 1988*, and to Walter Johnson and Julianne Meyer for the wealth of knowledge they recorded from conversations with Peninsula residents in the 1980's and 1990's, published in their book, *Memories Hidden, Memories Found on the Old Mission Peninsula*.

I hope all the rest who supported and gave me their time will find their names in the stories in the book. I thank you all from the bottom of my heart!

Sincerely,

Evelyn Johnson

INTRODUCTION

Barns of Old Mission Peninsula and Their Stories depicts the history and character of a place through a portal of essential elements on the landscape.

Old Mission Peninsula is cherished by residents and tourists alike. This book contributes a documentation of traditional barns and a sharing of human scale stories and experiences—a rich heritage of the people and places of Old Mission Peninsula. Strengthening that presence of the past adds to a more balanced understanding of this vibrant area of our state today. Readers will find variety here: a casual history of early settlement, reminiscences of individual and family experiences, the evolution of farming practices, and a survey and physical descriptions of the barns of Mission Peninsula.

This 18 mile long finger of land protruding into Grand Traverse Bay was a point of early settlement for this region. Before 1839 when Reverend Dougherty came as a missionary, the unique character of this land and its climate attracted Native Americans. These earliest inhabitants utilized and respected the land, the water and the woods of this place. Following Dougherty, the families who settled Old Mission Peninsula built homes, barns, roads and communities to live and farm and work here.

Researcher and author Evelyn Johnson came to the Traverse area first in 1967 as a young educator, wife and mother. Her family returned in summer and spent increasing time here until Evelyn and husband Carl retired to the area and settled on Old Mission Peninsula. Evelyn's attraction to the barns on the landscape, her intrigue with history and enthusiasm for people and their stories have come together in this book.

Barns of Old Mission Peninsula and Their Stories is a part of an increasing number of regionally researched and published books. The research for this book combines oral history techniques and photographic documentation with traditional literary research methods. It is unique for the regional and personal stories it contains and because it surveys—not samples—the barns of a particular area. Old Mission Peninsula is an isolated geographic area important because of specific historical and contemporary issues of settlement and land use. Barns are a part of what makes the landscape of Old Mission Peninsula unique.

Increasing scholarly attention is being given to rural and vernacular history and documentation. *Barns of Old Mission Peninsula and Their Stories*…will add to that body of literature.

Traditional barns were and continue to be working structures at the heart of a farmstead within an agricultural landscape. In this work, the barns of Mission Peninsula are considered as a part of a larger context within the upper Midwest. These are American barns, typically built by second- or third-generation North Americans who came to settle from other parts of the continent. In these cases and in this time, barn construction and characteristics did not generally reflect the ethnic heritage of their builders or owners—except in isolated situations where migration came directly from European origins. Researchers are working to devise standardized simple and direct visual descriptive terms to convey the external features of barns. Most people see barns from an external view, while driving along the countryside, as a part of the landscape.

The exceptional nature of this landscape and the climate of Old Mission Peninsula remain an inspiration today. In 1994 residents established the first legal mechanism in the nation to selectively preserve the right to farm and to protect the character, landscape and view of Old Mission Peninsula. Purchase of Development Rights (PDR) and land conservation processes are crucial tools that are helping to preserve this place and the beauty of Old Mission Peninsula.

Barns of Old Mission Peninsula and Their Stories will leave you with a challenge. Observe carefully what is around you, both in the built and the natural environment. Everything has a story embedded within. Everything has a history behind it. Learning and knowing will inform what we do as individuals and as communities. The history and the lessons of our past are important to understand and to share. Our heritage can both inspire and ground our children and grandchildren, preparing them for their future as informed and engaged citizens. Awareness and knowledge can be a foundation for shaping our communities and regions and determining the way we will live in them and protect them. These barns and stories contribute to our lives in cultural, economic and aesthetic ways.

Julie A. Avery, Curator of Rural Life and Culture
Michigan State University Museum

MY PERSPECTIVE OF THE BARNS AND OLD MISSION PENINSULA

As I drive the roads of Old Mission Peninsula each day I see barns! There are barns by the side of the road, off the beaten path, out in the middle of a field, and nestled among large, new homes on the fringes of subdivisions.

From my perspective, I am interested in how these barns fit into the picture when reflecting on the history of the area. Barns are living testimonials of our agricultural past and of the pioneers who were willing to come to put down roots and make a life and hopefully a living off the land. I want others, especially my grandchildren, to know that a part of what we are today comes from someone whose roots were firmly planted on a farm. I know that the Peninsula had twice as many barns once upon a time than we see today. I hope the next generation visiting Old Mission Peninsula will say, "I remember that barn!" instead of "Where have all those barns gone I used to see?"

We call this change…progress…but progress without the past is not real progress at all. We need to preserve what is important and I feel that part of that broad spectrum of history involves placing more value on our barns. So, in the summer of 2002, I gathered up my courage and started talking to the owners of barns on the Peninsula. My desire is to write the history down and take pictures before something happens to them. Because, you see, not only are barns pleasing to look at, but each holds a story and, if you will, secrets. They just need to be told and recorded. The information I have gathered from the willing barn owners keeps me grounded in the fact that the barns of our area *are* indeed where our cultural heritage comes from. In fact the word agriculture has the word *culture* in it!

The land on Old Mission started opening up in the mid 1800's for those of pioneering spirit. They came from the eastern states, from Canada, and from Europe to stake their claims. The soil was a dry sandy loam, and the land was heavily forested. The settlers, together, shared the work. It was remarkable that they could make a living, but they came to farm. The trees needed to be logged off. (Mr. Hannah and Mr. Lay had a lot to do with that. They bought the timber for their sawmill operation in Traverse City.) The stumps had to be removed as did the rocks and boulders the glacier left behind. The pioneers went to work raising crops, cattle and large families, and the area grew.

And, of course, they built barns. Around 1900 hemlock, pine, cedar, and other species of trees were logged off the property to build timber frame barns. These barns could easily have one long beam, thirty or forty feet long, often hand hewn, running the width/length of the barn on both sides. This confirms how huge many of the trees were on the Peninsula. The foundations were constructed of fieldstone as it was so prevalent. Barns originally served as farm factories, places of business, and storehouses for equipment and supplies. Farmers threshed grain on barn floors, stored hay in the lofts for feeding livestock, and sheltered animals in the stalls. The barns were the heart of the farm and were often built before the house.

Another type of barn found in the area was the plank frame barn, using sawn lumber from local saw mills on the Peninsula or from Traverse City. Usually the sides were constructed on the ground and raised into place. These barns started showing up in the 1920's and 30's. For practical reasons planking on most barns was nailed in place vertically for ease of construction and to shed water more readily. The long vertical planks were nailed side-by-side tightly. However, as the wood dried, the siding would shrink and helped to provide ventilation, which in turn helped reduce spontaneous combustion, which caused many a barn to catch fire and burn. Battens (narrow boards to seal the gaps between the large boards) were used in parts of the barn where ventilation was not an issue. These were the two basic types of barns built on the Peninsula.

It seems most of the local barns are bank barns…built into a hillside to utilize the lower level as a place to house the cattle. The barn would have one or two levels above. There was either a bank or a ramp for entering the main floor. Most barns, I've observed, have a bay on either side of the main floor and thus were called three-bay barns. The roof line was often the familiar gable end. The gambrel roof barn, whose purpose was to make the most use of the space within the loft area to store hay, was the other type of roof widely used. As need arose for more storage, many barns added three-wall extensions. These held farm equipment and tractors.

There were once about six distinct communities along the 23 miles of the Peninsula…often where the schools were. There were ball games, box socials, picnics, community dances, and many other activities in these communities. Now I think we have only two community and social centers. One is the Old Mission School that houses the Peninsula Library. The other is the American Legion Hall farther out on Swaney Road.

When fruit began to be the major crop grown on the Peninsula, many migrants came from Texas, Mississippi, Louisiana, and Florida to pick the crops…mainly cherries, apples, grapes, and other fruit. This was in the 30's and 40's. During the Second World War when at least 10% of the local men were serving in the armed forces, Mexican nationals were brought up. Jamaicans and Puerto Ricans were used also. An interesting note that I found in the *Greetings from the Peninsula Homefolk* book, which was a newsletter published for the servicemen by Old Mission residents, said "the picker situation is somewhat different than other years for there is a group of Japanese-American boys and some older folks to chaperone them…about 100 in all" (July-August 1944). In another *Peninsula News*, Frank Phelps wrote: "Last cherry time 50-60 girls and young women were quartered in the Community Hall at Old Mission, as part of the Women's Land Army. They did not rate high as pickers, but they were not too hard on the eyes" (March 1, 1944).

By all reports the migrants were pleased with their work and surroundings well into the 1960's. Early on they lived in tents or in unused barns. Later, housing units were built for their use. Eventually the federal government stepped in and tried to make the living standards of the migrants better. They wanted them to have hot and cold running water, screens on the windows, and more pleasant living conditions. Along with these demands, the government had other stringent guidelines to be followed. It became too expensive. About then, technology took over.

The cherry picker, the single limb shaker, and later the trunk shaker completely changed the cherry picking scene. Norm Crampton remembers the early cherry shakers well. Leland Gore and he were among if not the first to buy the limb shaker around 1968. He had one, called a Friday, and it was mounted on the front of a Ford tractor. As Norm put it, "It did the bull work." It was looked on with much skepticism by neighboring cherry farmers, but it worked! Other cherry shakers by name were the Shipley, the Perry, and the Arnold K. White. There were many attempts to construct a working shaker from parts of existing farms, some with greater success than others. Though there is still hand picking today, most of the cherry harvesting is done by mechanical shakers.

Barns are disappearing at an alarming rate all over the nation. We have lost some as a result of the 1950's and 60's trend of decorating rooms with barn wood paneling. These barns were not valued by the farmland owners, and they were glad to get rid of them as they became a burden to keep up. It is sad that there are more than a few barns on the Peninsula that are in need of repair. First, the roof starts to go and then the barn is left open to the elements. Old barns get less use because modern machinery does not fit in the barn built to house cattle, horses, wagons, etc. Hay isn't needed because, for the most part, cattle are no longer kept. Thus the barns fall into disrepair from lack of use and neglect. You might say they have become something of a dinosaur and they are expensive to repair and insure! But I am happy to report that I have noticed at least six barns that are sporting a new bright coat of red paint this past year. For the love of old barns, I say, "Hooray!"

Next, how did Old Mission Peninsula gain popularity and population eight years before the first permanent settler came to Traverse City? This is worth telling.

A BRIEF HISTORY OF OLD MISSION PENINSULA

When Reverend Peter Dougherty and John Fleming rounded the point jutting out from the Old Mission Peninsula in May 1839, rowing their Mackinaw boat into the east arm of Grand Traverse Bay, they "saw a wild uncivilized stretch of beautiful forest, and landed in a little cove known now as Old Mission Harbor. It was the home of the Ottawa Indian and white man had not yet struck his axe into the virgin timber or disturbed the Indian's native haunts, when the Presbyterian missionaries came" (*The First Protestant Mission in the Grand Traverse Region*, Craker, p. 34). When Reverend Dougherty arrived, the history of farming on Old Mission Peninsula began. The Indians began farming, and the white man started coming to settle the land around 1845. Thus the community of Old Mission became one of the first white settlements south of the Straits of Mackinac in 1840.

By treaty, the government had promised to supply a blacksmith, a carpenter, and a farmer to help the Indians cultivate the soil and become "civilized." However, Reverend Dougherty and his helpers were sent out by the Presbyterian Board of Foreign Missions and worked without any help from the government, save for the allowance of medicine dispensed by its agents for the Indians. (Craker, p.59).

"An Indian Treaty made in Washington D.C. in 1836, by which the Ottawas and Chippewas secured reservations to their lands for five years, was unofficially extended indefinitely until Grand Traverse County was faced with a problem—what to do with the Indians. White settlers were coming in rapidly and the terms of the 1836 treaty had long since expired. The adoption of the revised state constitution of 1850 made citizens of all persons of Indian descent who were not a member of any tribe. Thus, the Indians could buy government land, settle upon it and become citizens. However, Old Mission Peninsula was not yet on the market for settlement. Reverend Dougherty advised several of the Indian families to reserve part of their annual government payment for the purchase of land" (*The Story of Old Mission*, E. Potter, p. 67, also *Sprague's History 1903*). The Indians eventually moved across the bay from Old Mission Peninsula, where they could buy land to farm. Reverend Dougherty went with his "flock" in 1852. The "New Mission" was later called Omena. Ruth Craker in her book, *The First Protestant Mission in the Grand Traverse Region*, gives this explanation for the name. "Many times the Indians came to Reverend Dougherty to ask a question or give him an interesting bit of information. He would reply O-Ma-Nah, meaning, "Is it so?" The Indians thought that this term would be an appropriate name for a settlement and so it was named" (Craker p. 86). This has been labeled as being 'folklore' since Ruth Craker's time and not necessarily true.

The settlers who stayed were essentially squatters until an act of Congress in 1853 granted "bounty land to certain officers and soldiers who have been engaged in the military service of the United States." Actually there became several sources of purchase for the early settlers: One, the Peninsula was still an Indian reservation and settlers could buy the Indians' "possessors rights" from the United States Government. Two, purchase land from the state of Michigan. The St. Mary's Falls Ship Canal Company was building the canal there and Michigan sold large tracts of land in northern Michigan to finance its construction. Three, squatter's rights of the wilderness, which meant you picked out a piece of uninhabited land and settled down on it until it came up for sale by the government (*The Story of Old Mission*, E.Potter, p. 67).

"The Homestead law, giving to every actual settler from 80 to 160 acres of land for a merely nominal sum, which took effect on the first of January 1863 contributed not a little to hasten the settlement of the country. In the spring of 1874, Indian reserves were opened to homesteaders and the scramble for choice locations was on. A report published in 1865 by the state geologist, A. Winchell, stated that this area was most favorable for the cultivation of fruit. Professor Winchell reported that it was doubtful whether any other part of the U.S. could compete with it. John Garland's apple orchard on the Peninsula was noted in the *Traverse City Herald* (1859) as good as the editor had ever tasted. The rest is history. By 1874 the reputation for premium orchards for its fruit was established and the triumph of its pioneer fruit growers was complete" (Leach, p. 146-7).

All eventually received title to their land, one way or another. More than one landowner showed me the original land grant title signed by Abraham Lincoln, given after the Civil War. An original land grant was 160 acres. But they were usually split up among the family members or sold off. The land grant owners had to stay and work their land for 5 years.

Many of the roads on the Peninsula have names of the early settlers….Gray Road, Wilson Road, Swaney Road, Nelson Road, Kroupa Road, Tompkins Road, Ladd Road, Brinkman Road, and Montague Road. Fred Dohm claims that many of the roads on the northern end of the Peninsula were named by the ladies of the Old Mission Women's Club in the 30's. The area was growing in numbers of residents and all the addresses were RR1. The Postal Service needed more specific addresses, thus the ladies assisted the Post Office by naming the roads after their own family names.

Peninsula Township Major Roads

THE BARNS OF OLD MISSION PENINSULA 1850-1900

Picture courtesy of Carol Lewis

The barn/house

The farmstead in 1950

The house

RHEIMHEIMER BARN/HOUSE AND VIDA HOUSE

This house, once a barn, was built in 1890. The house on the property dates from 1880 as one owner, the Lees, discovered newspapers dating 1880, (*Traverse Bay Daily Eagle*) in the walls when they were remodeling the house. The newspapers were being used for insulation.

The first owner, found thanks to Rick Vida's research, was Cornelius Hawkins whose name appears on the 1836 plat map. He had 35 acres. His name remained on the property until he sold it to Victor and Olive Scofield in August 1908. The Scofield name was still found on the 1930's plat map.

Physical Description
Date Built:
- 1890

Construction:
- *Timber frame on grade barn*
- *Rough sawn vertical wood siding*
- *Wood shake gable roof*
- *Cement foundation*

In 1950 George Munson bought the house, barn and 30 acres. The Munsons did not use it as a farm. The aerial picture of the property taken in 1950-55 shows no orchards or crops. Bob Munson, son of George, remembers growing up on the property. The field out in back of the barn was his playground. Up the hill was a woods of evergreens and hardwoods. There were about 20 pear trees between the barn and Center Road, being the extent of the farm's orchards. Bob said he played in the barn and loved to tinker with an old 1947 Ford truck he found there.

Bob remembered the barn very well. It was of timber frame construction, with a wooden shake gable roof and rough sawn vertical siding. It was on grade with a cement floor and large double door at the back of the barn. Inside the squared timbers were about twelve inches wide and were pegged; no nails were used that Bob can remember. There were three stalls on the main floor that were about twelve feet wide. Each end of the barn had a loft with stairs leading up to it. One stall was used as a goat pen.

In the late 1950's and early 1960's some of the long time residents of the Peninsula remember a barn sale put on every summer by the Archie Community Women's Club. It was held in the barn and was to benefit the migrant workers who came to the Peninsula during cherry season.

In 1968, a Mrs. Long bought the barn and used it as a house, living there year 'round for 4-5 years. Bob told me she was a bit of an eccentric living in a very "basic" style. Bob said his dad helped her put in water, by drilling a well, and tried to make it more livable, but it was still a barn.

In the late 1960's the Munsons sold off the acreage behind the house and barn, creating 25 lots called Council Oaks. The name came from five oak trees that were in a circle on the property behind the barn. They had grown together, and it was said the Indians held councils there. At one time this was reportedly the largest red oak known, according to Bob Munson. Thus the subdivision was given the name Council Oaks. There still is evidence of the huge tree, according to Rick Vida.

In the 1970's, the next owners, the Schultzes, turned the barn into a house. According to Brett Baird and Sue Sabot, who bought the barn/house in 1994, an architect in the Schultz family designed and partitioned off the barn, turning it into a comfortable four-bedroom, two bath house. It also has a loft (perhaps one of the originals), a living area, galley kitchen, and a two car garage. It is fully insulated and has smooth plastered walls with 1080 square feet of living space. In 2006 the house was sold to Dean and Alexandra Rheimheimer. The couple are re-doing the house and installing a real barn door in the interior to bring back the authenticity of the barn that it once was. A lovely modern house was built in between the 126 year old farm house and the 116 year old barn.

The farm house just south of the barn-house was beautifully restored by Tom and Libby Lee in 1994. Rick and Rosie Vida bought the house in 1998. This victorian style house with a wide front porch is trimmed in green and has green and white striped awnings. The front yard is enclosed with a white picket fence, complete with a gate. The house is a very pleasing sight on the landscape, and the Vida's view is most pleasing too, as they look out on East Bay!

Information provided by Brett Baird, Sue Sabot, Millie Shea, Rick Vida, Mr. and Mrs. Richard Shantz, Bob Munson, and Georgia Goodman.

Corner of Swaney Road and Center Road in 1930's, looking south

DENNIS BEE BARN

This barn, though no longer used or needed, never the less comes with a lot of history of the families who owned it. Let's start at the beginning.

In 1850, when John and Rosannah Swaney came to Old Mission Peninsula, there were just two other white families in the area and about 500 Indians. They had made the long trip from their home in Pennsylvania through the Great Lakes with nine children, animals and belongings. There were no docks at that time, so as they neared the shore, their animals were pushed overboard to make it to land on their own. Indians in canoes met the incoming ship to unload passengers and freight. The Swaneys built a primitive cabin from slab wood to shelter them their first winter and eventually a farmhouse near Swaney Lake. This information and some of the following came from the Swaney Family History compiled by George Beckett, 1982.

> **Physical Description**
> Date Built:
> * 1898-1900
> Construction:
> * Timber frame on grade barn
> * Gable roof with shed added in the 1950's
> * Metal roof
> * Stone foundation

Both John and his eldest son Chris claimed 160 acres each. The land in 1850 was not on the market, thus they had to wait until 1860 to officially buy it for $1.25 an acre. Later, the family built a farmhouse on the corner of Swaney and Center Road, the barn being built across Swaney Road to the north. When the elder Swaneys passed away, Rosannah in 1868 and John in 1870, the youngest son, James, obtained 80 acres and the homestead. In 1871 James married Harriet Lardie. Together they had four boys and three girls. After the home burned down, it was rebuilt. The original barn, built c.a. 1898-1900, is still standing.

Jack Swaney, eldest son of James and Harriet, took over the farm when his father died in 1919. Jack and his nephew Ted Ayers farmed the forty acres together. Jack had been given 20 acres from his father when he turned 21. Thus they had a 60 acre farm. Two other brothers, George and Lewis, also had 20 acres given to them by their father, James. Their property and Jacks were all along side each other on Center Road.

The barn is now beyond repair and needs to be torn down, according to the current owner Dennis Bee. But, he believes, "It has many good beams and barn wood siding that could be used. It has outlived its usefulness." The barn once housed hogs, 10-12 cows, and three horses whose names were Jane, Ray, and Rock, according to Tom Bee, Dennis Bee's father. (A related story follows.)

Ted Ayers married Gladys Bee in 1946 and came to live on the farm. With her came her nephew, 12-year-old Tom Bee, whom Gladys had adopted when his mother died. When Tom Bee joined the family he remembers the crops grown were corn, potatoes, hay and, of course, apples and cherries. Tom explained the laborious way they planted corn in the 1940's. "A length of chains would be dragged across the field by two people. Then they would drag them across in the other direction, making a grid. Next, each man had a hand corn planter which they would walk along the 'row' created by the chain marks in the earth, and plant corn at the intersection of each grid." Tom's job was to cultivate the corn.

Water came from Swaney Lake and from cisterns around the house and barn. It was well after World War II that electricity came far out on Old Mission Peninsula where the farm was. The Ayers/Swaney family always had a Delco battery-operated system to make electricity. Jack Solomonson said you had to have four houses (farms) in a one-mile stretch of road before the electric company would be willing to hook you up to electricity.

In the winter when the ice was thick on Swaney Lake, farmers in the area would work together cutting chunks of ice from the lake. They would push the ice up on the sleighs and the horses would haul it home to their ice houses. Everyone had an ice hous, before electricity came along. Tom Bee remembers the ice was packed in snow, and then in saw dust from the saw mills on the Peninsula. It kept through until the next season and was their only source of refrigeration.

Uncle Jack told Tom that in the winter time, they would go to Elk Rapids across frozen East Bay with horse and sleigh, rather than to Traverse City. The distance was shorter!

Dennis Bee is the most recent owner of the farm, having abouty 150 acres mainly in cherries. He is the son of Tom and Barbara Bee. Dennis's grandmother, Gladys Ayers, lived to be 93 years old. She stayed in the family home, and Dennis remembers Grandma bringing the milk over to the house across the road from the barn to separate it. Milk and cream were still being sold to local residents and resorters alike around 1975. Cattle have not been raised on the Peninsula for a good long while. Dennis got his first cherry shaker in 1971-2. He sold about 60 acres of land to pay inheritance taxes on the farm when he took it over. The Peninsula township purchased a conservation easment in May 2005, on a ten acre parcel containing the barn. Dennis has farmed for twenty-seven years, and uses a large pole building that is on his property for his equipment storage now. The old barn and its occupants hold fond memories from the past. A story from the past follows.

Oral information provided by Tom Bee, Dennis Bee, and Jack Solomonson.

A STORY FROM MARY KEENAN
A former summer resident on Old Mission Peninsula, now a permanent resident of Traverse City.

Tawazi, an Indian word for "tall pines," was a wonderful cottage on the shore of West Bay. We rented this place the summers of 1960 through 1969. It was an old, large, and ramshackle place, but it had the most spectacular beach on the Bay surrounded by fragrant pine woods. It was perfect for a family with eight children, a large Labrador, and many visiting friends.

Our early teenaged daughter, Deirdre, whose birthday fell on August 26th, frequently reminded us that she wasn't able to have a birthday party with her numerous Detroit friends, even though we always celebrated it as a large and noisy family. On her birthday afternoon, we suggested that she and her younger brother and sister walk over to Lardie's General Store in the village of Old Mission to buy ice cream or candy. It was probably two miles away. Down shady Whispering Trail and dusty Swaney Road they meandered to the corner of Center Road, where they always stopped at Ted Ayer's barn.* It was the home of a gentle white horse who always seemed to enjoy the kids' visit…especially when they brought treats. The horse always seemed to be in the barnyard, perhaps hoping for a visit from his young friends.

On they walked down Swaney Road to Old Mission. Fortunately the proprietor, Bob DeVol, welcomed children and was patient with their indecision. Should they have a popsicle, pop, ice cream cones or candy? These were the days before healthful snacks were very important and necessary. Of course the journey home included another visit to the red barn and the white horse before heading back to Tawazi where the preparations for a birthday dinner were anticipated.

When they arrived at the cottage, they were amazed to find nobody and no preparations for the birthday celebration. They decided to run to the beach, down the juniper fringed path, with Deirdre leading the way. There they were greeted with balloons, "Happy Birthday Deirdre" signs, and a blasting record player attached to many extension cords trailing from the house. Shouts of Happy Birthday came from all the young beach residents and there was a fire pit with hot dogs and s'mores ready for consumption. Deirdre was totally surprised and thrilled to have a real birthday party complete with a spectacular sunset and the appearance of thousands of stars twinkling over the Bay. It was one of her very best birthdays!

*The barn on the corner of Swaney and Center Road belongs to Dennis Bee now.

Painted barns were often the color red, which was thought to have kept the dairy barn warmer in winter, absorbing the rays of the sun…when it did shine. Red is also a friendly, warm color indicating that the farm family was neighborly to all who passed by. A white barn, on the other hand, was said to keep the livestock cooler inside the barn in hot weather.

DR. JAMES M. JOHNSON BARN

This classic old timber frame barn was built around 1880. It is 30 x 40 feet and has an attractive wood shake gable roof complimenting this old weathered barn.

The Bell family homesteaded this farm in 1841 and owned a large amount of land on the Old Mission Peninsula. The barn was built by Erick Seaberg when he and his wife Jenny bought the property in 1887. The house was built behind the barn with the road running along East Bay. The Seabergs lived there raising their family and farming. They grew potatoes, corn, hay, had cattle and planted an apple orchard up on the hill behind the house. In 1910 Jenny Seaberg died. Erick married again, and he and his new wife Anna had one child, Opal. Anna was 45 and Erick was 65 years old at the time.

In the 1970's Neil Hanson bought the property, according to Dr. Johnson. Hanson had an extensive amount of frontage on the water with Bluff Road running along the edge of it. He changed Bluff Road to the other side of the barn, making it possible to sell off the lake frontage for home sites. The road now separates the house from the barn. By this time only part of the house, built in the 1880's, remained.

> *Physical Description*
> Date Built:
> - 1880
> Construction:
> - Timber frame barn
> - Vertical board and batten siding
> - Wood shingle gable roof
> - Stone foundation

In 1984, Max O'Neil was the owner of the property for about 10 years. His claim to fame is that he invented Kitty Litter. By this time the barn was in disrepair and was completely taken apart under the direction of Max, who was born on a farm in Iowa and loved old barns. The barn was rebuilt with all the original wood and it has been written into the deed that this barn must be preserved. It is a very pleasant sight to round the bend of Bluff Road and find this rustic old barn right there for all to see.

Inside the barn is a hay track, and the hay car for pulling hay to the top of the barn is still in place. At one time, I presume, it held quantities of hay. Today, it stands pretty much empty and is used for storage. The entry door is unique. It is constructed of slab wood and is very low. Most would have to duck down to get inside the barn.

*Information provided by
Dr. James M. Johnson
and Blossom Seaberg Jerue.*

CAUCHY/GRIFFIN BARN

From Jayne Griffin Boch I learned that Grandmother Griffin's husband died when she was pregnant with her fifth child. The Griffins already had three boys and a small girl when Mr. Griffin died of blood poisoning after stepping on a rusty nail. He had been employed by the Brush family on their farm near town. Mrs. Griffin and her husband had been in the process of buying some land and a house, and now the family was faced with a dilemma as to where they would live. The Warrens and the Buchans came to her aid, building a house for her on Gray Road, the Warrens being related to the Griffins. This tragedy happened around 1898. After the boys got a little older, Grandma went to work for a Mr. Ellis as a housekeeper, as he was a widower. The boys worked out too.

Probably the original owner of this property was a Mr. Willis Ayers who had a large tract of land (around 120 acres) on the Peninsula on the 1856-61 plat map. From information found by Charles Cauchy, the present owner, I learned that Thomas Gilmore came to Old Mission Peninsula in 1882 and purchased land from a Mr. Willis Ayers. The house was a shingle covered one story log house built by Mr. Ayers 35 years before, or in 1847. In 1888 Mr. Gilmore built a new frame house where the family lived until around 1900 when the family moved to Traverse City. The house still stands on the property.

Jayne Griffin Boch said her father, Dudley Griffin, worked for Guy Tompkins, as did the other children in the family. Dudley eventually bought 35 acres from Guy Tompkins, a part of what was the Gilmore farm, I am assuming. Jayne said her father was proud that he paid the farm off in a short period of time. The farm was deeded to Dudley Griffin in November 1909.

Physical Description
Date Built:
- Around 1880

Construction:
- Timber frame bank barn
- Vertical wood siding
- Metal gable roof
- Foundation: unknown

The large barn and shed with their weathered planks have never been painted and are plain and unadorned, except for the lightning rods on the roof. From the history of the farm, the barn was probably built around 1880. The gable roof barn acquired a metal roof in 1985. It is built into the side of a hill, making it a bank barn. "There were electric lights in the barn early on (between 1900 - 1920). Inside the main farmhouse basement was a wall of batteries and a Delco generator to charge them. In the barn there was a single cord and light bulb hanging from the ceiling. The hay wagon drove in from the back of the barn. Hay filled the top of the barn and made good insulation. The heat from the animals below kept the barn warm. The horses were kept on the right side of the barn. Flory and Babe were the team, and the other horses were Dick, a white horse, and Prince, a roan. These horses would have been used to pull a buggy in earlier times, and were kept on for the cultivator. The left side had stanchion for a few cows." Jayne has clear recollections of the barn and farm where she was raised. A silo was attached to the cow's side of the barn. A tractor would be hitched to the silage filler and it was shot up to the top of the silo. Jayne remembers the sweet smell of silage. Silage, she explained, is the green fodder made from corn shocks. There was a ladder on the inside of the silo and doors from top to the bottom. Silage would be pitched down, when needed, from the top all the length of the silo. You would use the door next to the silage level. Jayne said the cows loved the silage.

The Griffins had a self-sufficient farm. They grew mainly cherries, but also peaches, plums, and apples. They had a pasture and also grew hay. In the large shed near the barn Jayne said her dad made barrels for packing apples in earlier years. They sold them in town. Money would be borrowed in the spring to pay for fertilizer, seed, etc. and groceries were charged at Lardie's store in Old Mission. All was paid off after the crops were sold. The cream and egg money usually went to the wife for household items.

There were springs on the farm to help with the water situation and with the spraying rigs. Her dad attached pipes to the springs to water the cattle and other livestock. At one time there was a large stone and cement tank, about 10 x 10 feet, 4 feet deep and a foot thick to hold a water supply. When it wasn't being used for cherries and other things, Jayne remembers the kids had their own swimming pool in the barnyard, after it was cleaned and filled with fresh water.

Cherry pickers were picked up at the train station in town and consisted of drifters or out of work "rail riders." Jayne said her dad, when working and managing the pickers, always liked to wear a tie.

Gingerbread trim and lightning rods atop house built in 1880's

Dudley was the area road commissioner for a time. He helped improve Township Road, grading and graveling it. He also had a wooden V shaped snow plow pulled by a team of big horses. Jayne remembers seeing those big horses high stepping through the three foot drifts to pull that plow and snow out of the way, coming down Smokey Hollow Road, in the 1920's and 30's.

The roads were a local school district's responsibility back in the 20's and before. When you drove out of a school district, the road maintenance was the responsibility of the next school. Maurice DeGraw tells in *Memories Lost, Memories Found* (p. 205)," For a long time, it was like that (with the roads). Then they decided they should be combined and they elected road commissioners. A road commissioner was the most important man. They had a supervisor and a clerk, a treasurer, and trustees. A road commissioner had charge of building roads. That lasted way until the thirties when the Grand Traverse Road Commission took them over."

Dudley and Florence had three girls and one boy who died at two years. One daughter, Leah, was plagued with serious health problems but was always of cheerful disposition. She remained on the farm after her father died, and lived to be 90 years old. Jayne's mother, Florence, died when Jayne was nine months old. Dudley remarried Julie Persik from Leelanau Peninsula, and had two children, John and Betty. Betty was six years old when she was overcome by fumes from a gas pump and died.

After Leah died in 1972, the farm was deeded to John Griffin. He sold it to Phillips Energy. Now it is owned by Charles Cauchy. The barns stand idle now and are down below the road in a little valley behind the old farm house. To me, they are dark, old, and mysterious.

I am indebted to Jayne Griffin Boch for all the interesting information about the barn and growing up on the farm.
Also to Charles Cauchy, current owner, for the information he found written by Donna Hollister, granddaughter-in-law of Mr. Gilmore.

Plowing roads in the winter
W. Dohm, D. Eiman, D. Griffin, & R. Holmes

9

CHOWN BARN

This splendid gambrel roof barn was built in 1870 with the house being built about the same time. According to township records, the home was the fourteenth home built on the Old Mission Peninsula.

The original owners of the property were John and Della Marshall, the Sprague Biography says. "Twenty-eight acres were utilized for general purpose farming. John was active in township affairs, and his efforts on behalf of the general welfare have been far-reaching and beneficial."

Long-time Peninsula resident Molly Levin well remembered staying at Aunt Dell's when her family visited in the summer. Aunt Dell turned the three-story homestead into a boarding house/summer resort, and for four years Molly's mother, Helen Altenburg, ran the house with only one bathroom! The boarding house/resort served wonderful chicken dinners on Sundays and Molly's dad made ice cream twice a week, the old-fashioned way, with an ice cream churn. Molly said this still looms large in her memory.

> **Physical Description**
> Date Built:
> - 1870
>
> Construction:
> - Plank frame bank barn
> - Shingled gambrel roof, new
> - Shiplap siding, new
> - Fieldstone foundation

Molly remembered her family living in a little house out back of the barn. The barn was not in use at the time except for storage, though there were cherry orchards on the property. She remembered what fun she had with her brother throwing rocks at the barn one day. Of Course, quite a few of the windows didn't make it through the fun in one piece. The kids had to pay for them, according to Molly.

John Marshall's son Jules became the next occupant of the farm and he married an attractive school teacher named Grace, who was to have two more husbands (her subsequent husbands were Carl Pratt and Ed Bacon). Grace loved the farm and never wanted to leave it, but as her health deteriorated, she was forced to go to a home in Traverse City. Shortly before she died in 1986, she asked to be driven to the farm one last time to say goodbye, such was her attachment to the beautiful spot.

Grace had no heirs when she died, so the property was purchased by Jeanette Shambaugh Elliott, who had acreage down on West Bay on what is called "The Illini Strip." Jeannette was the last owner before the Chowns came on the scene.

Glen Chown credits his wife Rebecca with their good fortune in being able to buy the property. Glen runs the Grand Traverse Regional Land Conservancy, and he and Jeannette had worked together setting up conservation easements on properties Jeannette and her family owned. Shortly after Glen and Rebecca married, Jeannette invited them out to her home on The Strip for dinner so she could meet the bride.

This initial meeting in 1994 was a humdinger, and after an intense couple of hours in which Jeannette grilled Rebecca on her political, social, and educational views, she asked Rebecca if she'd like to see the farmhouse. "What farmhouse?" Rebecca responded. She had no idea what Jeannette was talking about, and Glen was equally mystified. A short while later, they were walking around inside the rambling home that had been lived in only sporadically for the past ten years.

The Chowns moved in on July 29, 1995, and have been working ever since to bring back the house, grounds, and barn to their original glory. The house is lovely and large and well on its way to being restored to a turn-of-the-century farmhouse, complete with cherry wallpaper on the dining room walls. The original maple floors have been uncovered and restored. Much of the rest of the house has also been renovated or restored.

The barn has been restored completely, mostly by Glen. When they bought the property, the east side was leaning badly, pulled down by a poorly constructed three-sided addition. They tore this addition down to disclose a beautiful stone foundation. Then they

Inside rafters of Chown barn

went to work shoring up and straightening the barn with tie bars to secure it. The whole thing was also reshingled, and the barn was sided with shiplap and painted gray to match the house.

Glen's and Rebecca's parents got into the action helping with everything from lugging new shingles up to the roof, to swinging a hammer on the scaffolding, and even picking up old shingles and nails on the ground. Various brothers-in-law and a few sympathetic neighbors helped as well. Rebecca thinks the new paint was the first the barn had seen in fifty to sixty years.

Glen learned how to side, roof, and complete all the other necessary work, mostly through trial and error, though he did have help in restoring the stone foundation by master mason Bill Love, who works for the National Park Service in Leelanau County. In addition Glen pounded out and re-poured cement on part of the barn floor, and did the same in the old milking parlor on the east side of the barn, turning it into a spacious stall for the two horses they have acquired, compliments of Rebecca's parents.

Glen and Rebecca are very proud of their farmstead restoration and feel a real connection to this land. They cannot count the hundreds of hours they have put into the restoring of the barn, but they estimate the total cost is somewhere in excess of $60,000.

One story that must be told is reported by Rebecca. When they first moved into the house, and for several years thereafter, strange things kept happening in the home. Household items would move around, strange noises would occur, and various lights as well as the stove would go on and off without any odd weather to account for the disturbance. Old timers on the Peninsula told them the house was haunted and that many tradesmen on the Peninsula refused to enter the home to work. Rebecca said she is not a brave person, so she was glad it was obvious the presence in the house was friendly, one that simply wanted to be noticed. Little by little as the Chowns improved the home, the frequency of the strange occurrences decreased.

One day as a neighbor's father was experimenting with a new metal detector at the Chowns, he found a gold wedding band buried in the gravel driveway. (Remember Grace was married three times.) The gold band fit Rebecca's finger perfectly, and after that there were no more signs of the ghost. The Chowns like to think that Grace is satisfied with them and can now rest in peace, since the love she felt for the farm is reflected by its current occupants.

Oral and written information obtained from Glen and Rebecca Chown, and oral information from Molly Levin.

Title to the Stone barn—1864

COTNER BARN

William Stone came to Old Mission in 1850 and was one of the earliest pioneers in the area. The Traverse City Land Office records verify that by 1859 he had been using his land for agricultural purposes for two years and had built "fences, a good barn and other out houses." He gained title to 58 and $^{90}/_{200}$'s acres on November 1, 1864. For a time after his arrival he was employed by the U.S. Government in the Indian department as a foreman. He later engaged in a number of very profitable business pursuits, including continued trading with the "red men on his own account." (Sprague Biography, p. 755-6)

He bought an early store down by the Bay from Lewis Miller in 1852 and had it moved up to the road. George Lardie bought the store sometime before 1881. Stone had become the postmaster in 1851 keeping the mail in a cigar box (some say it was a raisin box) in the back of the store. In fact, he was the only postmaster between north of Muskegon to south of Mackinac until one could be established in Grand Traverse (soon to be called Traverse City).

Physical Description
Date Built:
- 1859

Construction:
- Timber frame bank barn
- Vertical wood siding
- Steel gabel roof
- Fieldstone foundation

He drove the stagecoach between Old Mission and Traverse City bringing meat from Traverse City in his coach on certain days, and displaying it in trays in front of Lardie's store. "Mr. Stone has been prominent in public affairs and has left the impress of his individuality upon the development and progress in many ways. He was the county sheriff for three and a half years, during which time he took the first prisoner from Grand Traverse County to the state penitentiary at Jackson. He became one of the county commissioners of the poor and has been treasurer of Peninsula township." (Sprague Biography, p. 755-6)

Mr. Stone married Katie Corely of Northport. Several individuals remember they always called each other Dolly. They raised four children, F.W. Stone, who had land adjacent to his fathers, Mrs. John Holmes, Mrs. E.O. Ladd, and Mrs. Nellie Baldwin. Mr. Stone cultivated 65 acres of fruit trees, along with his many civic duties. Their 17 room home was open to paying guests and was listed as one of the four hotels and boarding houses in Old Mission in 1881. After his death at the age of 96, the farm passed to daughter and son-in-law Luther and Nellie Baldwin. Rebecca Tompkins-Nothstine, a long time resident of Old Mission Peninsula, boarded at the Baldwin house when she came to teach at the Old Mission school in 1938.

John Holmes married one of Mr. Stone's daughters but did not live on the property as he had a farm on Center Road. Much later, around 1960, Russell Holmes, grandson of John Holmes, lived on the property and Katherine, Russell's wife, had the old Stone house torn down to make room for a new house. The Cotners, Ronni and Vic, bought the property from Russell Holmes' son, Jack, in 1998. Ronni said, "We liked the barn and the tire swing and so we bought the farm."

This 147 year old timber frame barn is still standing. Vic and Ronni Cotner, the present owners, feel the barn has been repaired a number of times. The Cotners added a new steel roof on the barn soon after purchasing the property. The barn and other out buildings received a fresh coat of red paint. Ronni said, "The doors on the bottom level of the barn are pretty small for cows, but smaller farm animals or fowl were probably kept there."

The Cotners have 2.65 acres and use the land down behind the barn and house for a field of barley. The top of the barley is used to make mash for their horse, and the straw is distributed in the two ponds on their property to keep the algae down.

You can feel the passion in Ronni's voice when she talks about the ongoing project to improve and repair the barn which is used for storage now. They love the history and the mystery of what went on in this historic community.

Oral history provided by Ronni Cotner, Bob DeVol, and Rebecca Tompkins-Nothstine.

The Stone house picture by Carol Holmes.

The original William Stone house in Old Mission

FREDERIC and KAY DOHM BARN

Well worn shoes hanging on the side of the barn tell it all. One pair belonged to Frederic's father. Some of the others belonged to nephews of the Dohms who have worked on the farm. Looking at their condition, the shoes may be a reminder of just how hard working on a farm can be. I guess we could ask the Dohms.

Three generations of Dohms have inhabited this farm, referred to as "Twin Maples." The William Dohm family purchased the farm between 1920-22 from Miles Gilmore. At that time, cherries already were being grown on 20 acres of the land. Until 1890, when Miles Gilmore purchased the farm, it was a sheep operation. Along with many area farmers, Gilmore put in potatoes and had dairy cattle too. Across the Peninsula potatoes depleted the soil of necessary nutrients and thus there was a gradual change over to growing cherries in the early years of 1910-20. Fred, Sr. said it took a lot of work to bring the soil back to productivity after the potatoes.

Physical Description
Date Built:
- 1892

Construction:
- Timber frame bank barn
- Three wall shed added 1920's
- Shingled gable roof
- Fieldstone foundation

The existing barn was built in 1892. Miles Gilmore might have built this barn, as the initials M L G are carved inside the side door. Also there are the initials HW, HB, DE and JWK. Frederic Dohm, the third generation member of the Dohm family living on the farm in 2005, believes these initials may be the signatures of the crew that built the barn. Frederic remembers that in 1974 a more crudely built section of the barn was torn down. This gable roof barn is timber frame construction, has board and batten siding. There are many round log beams placed horizontally in the lower level of the barn. Some logs still have bark remaining. There is one log post in the center of the lower level. From the main floor you can see the roof rafters. In the 1920's a shed was added to the west side of the barn for machine storage.

Milking stanchions remain in the lower level of the barn where cattle were kept until about 1945, when the war ended, Frederic said. Built into the corner of the lower level is a poured cement water tank. Early on there was a wooden silo and a windmill that served this farm. These were common on many farms at one time. Frederic remembers the "sounds of windmills squeaking all along the Peninsula 30 years back," c.a. 1974. However, few windmills and silos remain on the Peninsula today. When not used, and maintained, they have a tendency to deteriorate rapidly. Fred Sr. remembers his father William joining with neighboring farmers to purchase a steam traction engine that powered a silo filler when there were more cattle on the Peninsula.

Throughout the Dohms' ownership this farm was largely self-sufficient as they always had a family garden. Frederic describes the barn as "the center of activity: animals to feed and care for, cows to milk, and a great many other work activities took place in the barn." Recreational and social activities were also held in the barn. One event—a cider party—is still hosted annually in the Dohm's barn. The barn now is used mainly for storage and is the home to many, many cats. Frederic put it painfully clear when he said, "Every product has its life and the barn has probably lived its life."

Fred Sr. (Frederick) and Frederic (his son) both attended Michigan State University. Frederic was so proud of his school that when State went to the 1986 Rose Bowl he celebrated by painting a big "S" on the front of the barn distinguishing it from other area barns. Frederic graduated from State and became a farmer, just as his father and grandfather did. Frederick attended State for one year. He joined the Navy when World War II needed a great many men in the armed forces. He did not wait to be drafted.

Migrant workers were crucial to cherry harvesting. Up until cherry shakers came into use in the 60's and 70's, the Dohms had 35-50 migrants come annually for the harvest. They were housed in five buildings on the farm, including the barn. Close relationships emerged between farmers and migrant workers. Workers were often loyal to one family and became good friends with the fruit growers. Fred Sr. remembers one Hispanic worker called him from Texas and asked for a loan to travel north to the farm to work. Fred didn't have extra money to spare at the time and referred him to the local bank who loaned money for the trip. When the season was finished, Fred laughingly remembers the fellow paid back the loan and brought the banker a lug of cherries as a bonus! The local sheriff had also helped the workers get up north and he received a gift of cherries too! One year, Fred Sr. and his wife went to Texas to visit their Hispanic friends, showing how well they thought of them.

Early photo of the Dohm barn

FREDERIC DOHM BARN

Frederic's great-grandfather, Charles Kroupa Sr., was one of the original Kroupas on the Peninsula. He was the only one of his brothers to go to the Civil War. He participated in Sherman's March to the Sea, helping to end the Civil War. Kroupa came home after 1865 with his feet "crippled" from the experience. He was regarded as a good farmer and also known as a wood carver. Frederic has a special family treasure, an alligator, carved by his great-grandfather. It is believed Charles wanted his children and grandchildren to see what an alligator looked like. He had probably seen them during the war and this was a way to show them.

Charles won a prize at the first Grand Traverse County Fair in 1878 for his wood carving of an almost life size crucifixion (October 10, 1879 *Traverse City Herald*). It is said he "could carve anything," and loved to carve animals. Charles Kroupa was born in 1833 and died in 1901, when he was 69 years old.

Information and pictures provided by Frederic Dohm and further information by his father Fred.

There are upwards of 16 cherry stops along Center Road selling cherries during July.

FLACHSMAN BARN

When William and Sherry Flachsman bought the property in 1994 they completely renovated the house, but the barn remains the same as when they bought it— an unassuming red hay barn with stables is nestled in among the apple and cherry orchards.

The barn was built by Peter Zoulek in 1870. Zoulek passed his farm on to his daughter Ann. Ann married a Kauer. Louis Kauer was the last of his family to own the farm. He died in 1986 in his 80's. Louie was an interesting character who loved horses and farmed with them all of his life. He also knew all the barns in the area. He never married and the farm went to Tom McManus who had cared for Louie in his later years. At that time the farm was 60 acres; today it is just five and one half acres.

This L-shaped, timber frame barn is on-grade. One of the L's held four horse stalls where Belgiun draft horses were once stabled; the other L has three bays with a threshing floor in the middle. Hay storage and lofts are on each side. Each side of the main floor is a good deal lower and dirt covered. Part of the gable roof is covered with corrugated metal and part of it is shingled. William (Woody) called it a hay barn—"no basement—no cattle—just for horses, and hay storage." He feels there have been many "fixings" and alterations to the barn over its long life. Both the barn and house foundations are large rocks and an occasional tree stump.

At one end of one of the L's is a spirit hole to let the bad spirits out and the good ones in. This is a superstition that came with the early settlers from their native country.

Oral information provided by William Flachsman.

Physical Description
Date Built:
- c.a. 1870

Construction:
- Timber frame on grade
- L-shaped with partial dirt floor
- Gable roofline part corrugated metal and part shingled
- Foundation building is setting on large rocks

HERR BARN

With a wide green lawn going down to Peninsula Drive and the Bay beyond, this little (20 x 28 foot) barn makes a postcard perfect shot. Its past owners were the Holcombs, Fragels, and the Augers. In 1970-72, Edgar and Grace Auger completely restored the house on the farm. The farm was unoccupied at the time, and it consisted of 20 acres of cherry orchards with 700 feet of beach front. Today the remaining farmstead of four acres is owned by the Herr family. The rest of the land was sold to develop Bayside Woods subdivision.

Built in 1898, this timber frame red barn trimmed in traditional white is now a storage barn. It is a bank barn, set back into the hillside with a lovely fieldstone foundation and handsome large beams supporting the upper floor.

When I asked Jerry Auger if any animals were kept in the barn he replied, "Only a 1965 Mustang convertible which we restored." Most likely it did hold animals—cattle and a horse or two in its earlier years.

Oral information provided by Jerry Auger.

Physical Description
Date Built:
- 1898

Construction:
- Small timber frame bank barn
- Vertical wood siding
- Asphalt shingled gable roof
- Fieldstone foundation

The social event of the season was the dancing party given by the girls of Old Mission on Wednesday evening. The finest dancing hall in the place was engaged for the occasion and was beautifully decorated with American beauty roses and ferns. The real beauty of the affair, however, was the girls who, with their happy young faces and pretty gowns, made the occasion one that will be long remembered.

JOHNSON/JOHNSON BARN

This large old barn is constructed of many different kinds of wood—whatever the owner could find to build with. In fact, the house on the property, namely the windows and hard wood floors, are from the Odgensburg Church when the original was torn down in the 1920's or 30's. The barn has curved window wells to let more light in down in the basement level. It housed cows and other animals. This barn would be just another gable end red barn if not for the truly unique shiny, ornamental ventilator at the peak on its metal roof.

Ernie Umlor, who owned the barn for many years, was a unique individual, according to Heather Johnson Reamer. He was an experimentor, an inventor, and an entrepreneur. He was very frugal. He could not read or write, but he could save money.

This 80 acre farm was considered poor for crops such as corn. Dean Johnson, who now owns the farm with Teri Johnson, remembers it was just scrub brush and the terrain was hilly. But when Mr. Umlor planted cherry trees he made money. Dean said, "Actually, he made more than my dad on the farm right next door, because he had more plantable acres."

There is an orchard to the south of the barn that was planted by Mr. Umlor. He decided to experiment with nut trees and found English and black walnut trees grew well. There were walnut shells all over the main/ground floor of the barn. The squirrels are the culprits. There are nut drying racks on this level that were designed by Mr. Umlor.

Physical Description
Date Built:
- 1880

Construction:
- Timber frame bank barn
- Vertical wood siding
- Metal gable roof
- Fieldstone foundation

This barn and acreage is reported to have been traded, with an ox thrown in, for the present Oliver Lardie homestead a mile or so down the road to the south.

A former neighbor, Mrs. Roger Kitchen, remembers, "Ernie had a pool table in another building on the farm that became quite popular with the locals who would visit his pool hall." Later, when the farm was acquired by the Johnsons, the pool table was donated to the Legion Hall in Old Mission.

Cal Jamieson remembers that Barry Loomis and he took Barry's mother's new sheets, made parachutes from them, and jumped off the roof of the Umlor barn when they were kids. They had been quite impressed by a local outdoor movie they had seen, presumably a war movie showing pilots baling out of their airplanes. They decided it didn't look too hard, if maybe just a bit risky. They are here today, Cal said, because they landed in a manure pile up to their armpits.

Oral information provided by Dean Johnson, Heather Reamer, Leona Kitchen, and Cal Jamieson.

19

Kelley's wellhouse

Original sawn timber planks
of varying widths inside the barn

The fruit buyers of the city are receiving cherries in considerable quantities, a full carload being shipped out last Saturday afternoon by George W. Lardie. The advertised price was 80 cents a crate which is considered a good price for early sour cherris.

KELLEY BARN

H.K. Brinkman original log cabin in the 1850's

H.K. Brinkman, a self-made man, came from Prussia in May 1853 to Old Mission where he followed his trade as a shoemaker. According to *Sprague's Biography*, "After coming to Old Mission he continued to engage in shoemaking for five years. This was virtually the first manufacturing establishment in the Grand Traverse region. After living there five years he turned his attention to agricultural pursuits and purchased two hundred acres of land in Peninsula township. He had previously entered this as a squatter's claim but it had not regularly come into his possession. His specialty was apples and he raised many kinds and marketed and shipped them out of Old Mission Harbor. In 1899 he started a plant for the drying of apples and was most successful." He, by this time, had left the farming to his son Eugene. He also served as township clerk, treasurer of the township, and in 1864, served in the Union army. He was active in politics too. He still owns this property and now has about one hundred and forty acres of improved land, seventy-five acres being included within his orchards, raising splendid fruits—apple, peach, and pear orchards and also a large vineyard.

The barn itself was built in 1861 and was framed by S.E. Wait, an early builder in Old Mission. The large 22 room house was built the year before in 1860, and was called a double house. H.K. Brinkman lived there with his wife, Keziah, and four children, and his mother-in-law, Eliza Elizabeth Colburn. The family was prominent and well respected in the community.

In 1927, the house burned and was replaced later with the present home. This house was built by Oliver Brinkman for the hired men to live in while they took care of the farm. However, at the time of the house burning, the barn was left standing, and a lovely little stone building used for a variety of things, probably for cold storage, according to Paula Kelley. There was a log cabin on the corner of the property that was Mr. Brinkman's workshop for making shoes, but he also made shoes of a different kind, as he was a blacksmith.

Physical Description
Date Built:
- 1861

Construction:
- Timber frame
- Ribbed steel gable roof
- Shiplap siding
- Fieldstone foundation

The barn is still in excellent condition due to the care taken by the Kelley family that now resides on the property. Its original use would have been to house the cattle and the horses. The cow gutter is still in existence. The sawn lumber used to side the barn is of random widths—some two feet wide and some much narrower. The barn has a sturdy fieldstone foundation. It was recently reroofed in green ribbed steel, hiding the wooden shingle roof underneath which was beginning to leak.

Jackie Burns, who is a granddaughter to Oliver Brinkman, was sent a newspaper article by a relative. The name of the local newspaper is not known, as Traverse City had several newspapers 100 years ago. However, in 1986 this article had appeared in the newspaper in 1886." Mr. Tompkins, the father of several of the neighboring farmers, while passing through a field owned by Mr. Brinkman, was attacked by a ram. He was knocked down and badly used up. Because he is an old man, his injuries may be serious." According to one story told by Jackie Burns, H.K. Brinkman put a ram in his plum orchard to ward off the Tompkins who H.K. claimed were stealing his plums as they walked from their farm to "the dock" at Haserot Beach and selling them to the ships that came in. H.K. wanted to put a stop to that.

The farm eventually consisted of 325 acres due to various acquisitions. The farm was in the family for 100 years and was sold in the 1970's to the Oxtebys. Oxtebys sold a small parcel (about 6 ½ acres) of the 325 acres to the Kelleys about 2 years ago. The larger portion of H.K. Brinkman's original land is in a land conservancy and can never be developed.

I enjoyed a story from Bob DeVol told by Jackie Burns: "During the week, the baseball field (on Brinkman property) was a cow pasture and on the weekends a baseball field. It made for an interesting field to play on, especially watching the town boys slip and slide on what they thought was a base."

Oral information obtained from Jackie Burns and the Kelleys.

JIM RICHARDS BARN

This little barn is old and probably one of the first on the Peninsula. Jim Richards thinks it could have been built along with the house in the late 1840's. It is a timber frame, gable roof barn that is resting on large tree trunks. It has a dark, weathered look with just a hint of a long-ago red painting. Jim said it had to be reroofed with asphalt roofing, covering up the cedar shingles that were inches deep with moss. The barn has a loft and was partitioned off for horses and cattle.

The Richards' fifteen acres is deep in the woods and they can see both bays from a high part of the ridge they live on. The Richards call their property Ridge Wood. While the house has received a good deal of renovation, dating from 1967 when Jim's father bought the property, the barn remains as it always was.

Physical Description
Date Built:
- Around mid 1800's

Construction:
- Timber frame on log stump foundation
- Vertical wood siding
- Asphalt shingle gable roof

The property was settled by Captain McKinley who received a land grant from President Zachary Taylor for service in war, probably the War of 1812. His expertise was as a topographer. (surveyor). He had lost a leg in the war but moved to the northern part of the Peninsula where his land was, and presumably built the house and barn. When the Civil War began he enlisted again! At the end of this war, he received a land grant from President Lincoln. He is said to have come home a hero. From the Abstract of Title, I gleaned this: August 15, 1861, James McKinley, pursuant to an Act of Congress, approved March 3, 1855, entitled, "An act in Addition to certain acts granting bounty land to certain officers and soldiers who have been engaged in the military service of the United States." (The government survey shows the land contained 50.75 acres.)

In 1871 the Deed shows Joseph Archer as owner of about 25 acres of property. The other 25 acres were purchased by a Thomas T. Reese. Jim Richards, the current owner of the Archer property, said Mr. Archer was a botanist who experimented with grafting of fruit and is directly responsible for the fine tasting cherries we have today.

Alice Archer, daughter of Joseph, married T.T. Reese. In 1908 Alice Archer Reese deeded the land to her children, Ida Reese and Charles Reese, and Robert and Martha Belle Buell. Each received 1/3 of the property. The Buells ended up with the land in 1923 with Charles and Ida signing off. The house on the property became a residence for hired help on the Reese/Buell farm. When Robert Buell died in 1925 at the age of 46, Belle Buell married again. Her new husband was one of the hired men, William Welhusen. Willy was 21 and Belle was 48. Belle died in 1968 at the age of 82 and the property went to William and Alice Buell Thiel, Belle's daughter from her marriage to Robert.

Jim Richards delights in the history of the area and showed me the name "Harry Britten" stenciled on the inside of the barn. Harry was a hired hand who often slept in the barn. There is no doubt that this land was once inhabited by the Chippewa and Ottawa Indians as Jim has trail markers on his property, limbs that have been permanently bent to guide those that traveled along the path. Jim also thinks there is evidence that there was a cider mill on the property at one time, built after the white man came to live on the land.

The Richards family moved to the property in 1995 and are owners and operators of Old Mission General Store, selling everything from pickles, peanuts, and crackers in a barrel to hand dipped ice cream cones, gourmet foods, and souvenirs. What I like best is that Jim and Marcy sell Coca Cola in a glass bottle. With a purchase you could easily get an interesting tidbit of history from Jim Richards himself, as he loves the lore of this historic old community.

Oral information provided by Jim Richards.

SHULTZ BARN

The Shultzes, Gerald and Mary, purchased this on-grade gable roof barn in 1960. It is 32 x 46 feet and was built in 1880. It was formerly part of the Buchan farm and was used for general farming. The original farm was 65 acres, but was reduced to 27 acres in 1973 when a subdivision was built along the south side the the farm. The Buchans housed cattle in the barn. When the Shultzes bought the farm they had cattle for one more year. Now the barn is deteriorating due to lack of use.

The Buchans planted their apple trees on the steep hill in back of their house in a "contour" fashion, so that the orchards could be maintained with horses. Apples were packed in barrels and sent out of Bower's Harbor to Chicago. A big "rich" apple buyer came to the farm to look the apples over once, and lost his large diamond ring in the orchard. It was never found.

The Shultz house on the property has a striking stone exterior and looks out over West Bay. With the stoney fields and West Bay being bordered with rocks, it is no wonder the lovely house is beautifully built with stone. Mrs. Shultz pointed out that there are a number of houses with stone exteriors on the Peninsula. It seems clear that the early folks used the materials at hand.

Oral information provided by Mary Shultz.

Physical Description
Date Built:
- 1880

Construction:
- Timber frame on grade barn
- Vertical wood siding
- Metal gable roof
- Barn is sitting on boulders

Corner foundation boulder of Shultz barn

Shultz's stone house

DANA WELLS BARN

Dana Wells has always been a farmer. This 31 acre farm was originally owned by his parents, Floyd and Helen Wells. The barn is 30 x 40 feet, has horizontal siding and was built in 1880. It has a gambrel roof line with a metal roof.. It is painted a soft blue. It is a bank barn with the big barn door at the back of the barn so the animals could enter the lower level. It is not a big barn but has a certain charm in the way it is built.

In the 1930's and 40's the Wells grew corn and hay and raised feeder cattle and hogs. The milk cows were for their own use. The barn also housed a horse, and they had chickens. Cherries came next. Apples and prunes were also grown on the farm. The need for pickers prompted the Wells to build five outbuildings to be used as housing for the migrants that came from Texas and Mexico. Dana believes that the housing they provided for the migrants was a good deal better than most of them came from in the South. The Wells put in dividers to make apartments during the picking season. They utilized about sixty migrants. Dana said they would arrive in the back of a big flat bed truck with sides. A tarp was pulled over the top for protection. Much of the home grown produce that they raised was sold to the pickers. They also had eggs and milk for them.

Physical Description
Date Built:
- *1880*

Construction:
- *Timber frame bank barn*
- *Horizontal wood siding*
- *Ribbed steel gambrel roof*
- *Cement block foundation*

In the earlier days spraying the cherries was done by hand. It took three people to get the job done. One would drive the tractor and two persons held hand sprayers that resembled a gun. These two would have a sprayer in each hand.

In the 1970's when cherry shakers came to be used, the migrant labor was no longer needed. Dana said the cherry buckets and ladders from the earlier days are still in the barn.

Now that the Wells retired and are out of the farming business (since 2000), the barn is well maintained and used just for storage. Lining the driveway to the Wells' house is a row of beautiful maple trees which must be over 100 years old. Next to the driveway is an old, neat one car garage sitting close to Center Road. It was probably placed there because the folks could get out easily if the road was plowed, since driveways were shoveled out by hand.

Oral information provided by Dana Wells

Doc Jamieson and Bud Jamieson spraying in the 1920's. Garrett Boursaw is driving. Photo courtesy of Cal Jamieson.

Well's garage

THE BARNS OF OLD MISSION PENINSULA AROUND THE TURN OF THE TWENTIETH CENTURY

The tunnel of trees on Smokey Hollow Road

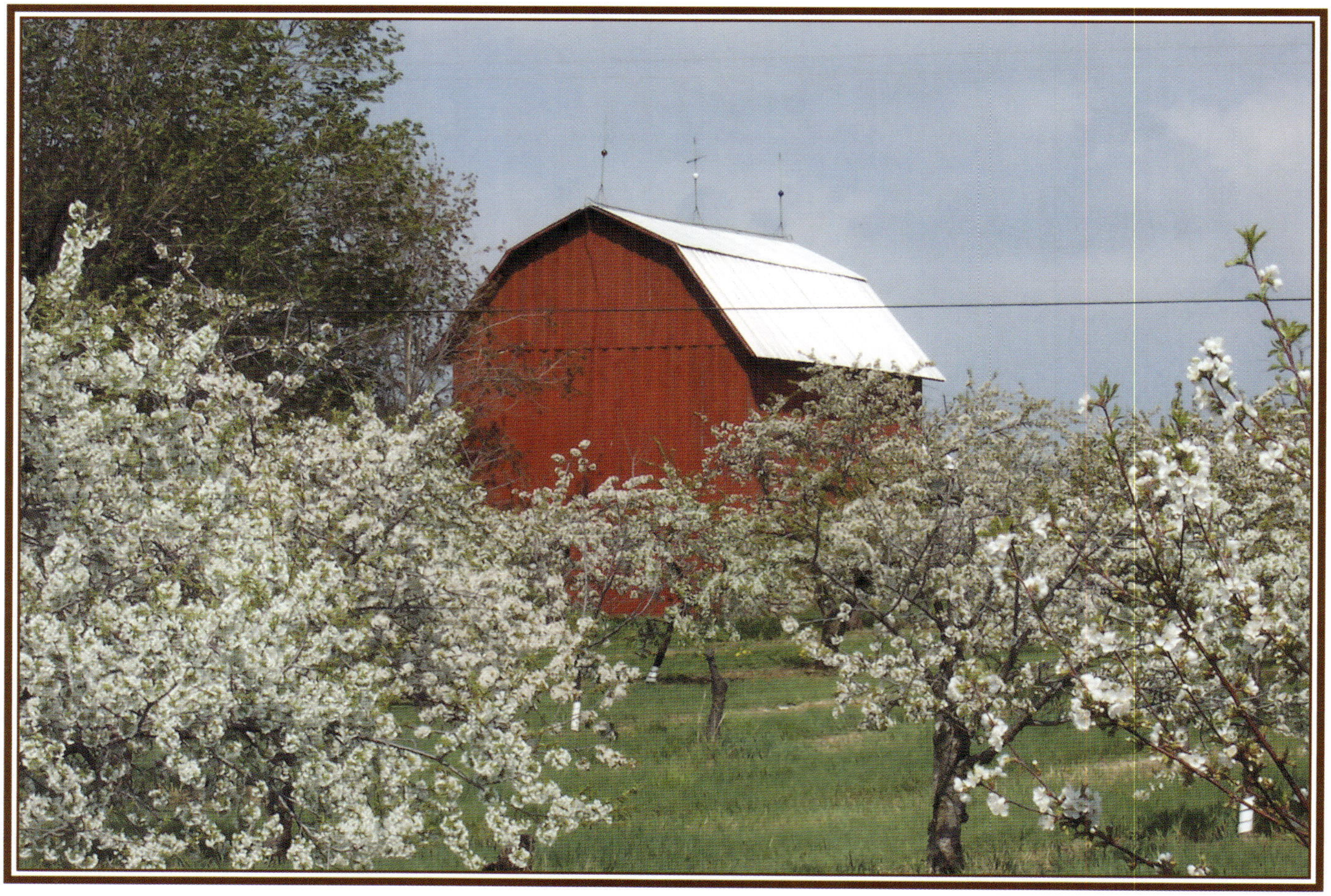

FAYE/ELTON DOHM BARN

Elton Dohm remembers the farm was a lot of work, but I doubt any Peninsula farmer would argue with that. I talked to Fred Dohm who was a cousin of Elton's father, Faye Dohm. He said Faye was a very hard worker, was thrifty and kept an immaculate farm setting. Elton lost his father when he was in his 80's.

This gambrel metal roof barn was built in 1900 and still has its lightning rods in place on the roof. It has been maintained and the farm is rented out now. There are approximately 10 acres of land and some cherries are still harvested from the land. The original farm was 22.2 acres.

Cattle and a team of horses were kept in the lower level at one time. Corn and grain were grown for the animals and, according to Elton Dohm, the family always had a big garden. Elton said his mother purchased extras in town from the cream money.

Fred, who is 94 (2006), reminisced about days gone by, and told me many good stories. Here is one I particularly enjoyed. A group of locals from the Peninsula went deer hunting each year and thoroughly enjoyed themselves. They pitched a 16 x 12 foot tent and really roughed it, camping eight miles off the main road. They kept the November chill out of the tent with a barrel stove set up on rocks. They hunted all day and did not come into camp until evening. One year they all had their pictures taken with six bucks hanging from the buck pole. The next day they got one more buck just before they broke camp and were able to get out of the Upper Peninsula ahead of a big snowstorm. The picture shows the seven gentlemen in front of the deer, the photographer being John Langworthy, an early settler on the Peninsula. An outspoken man, one evening in camp he expounded on how to raise children. He had no children of his own, but he did have a lovely collie. He had very definite ideas and was giving endless advice to the other hunters who all had children of their own. Finally Faye, father of two, quietly said, "If you had a lively boy instead of that long haired dog, you'd know what you were talking about!" The conversation by Mr. Langworthy was over! These men hunted together for 40 or so years, and I am sure there were a few more good stories to be told.

Information from Elton Dohm and Fred Dohm.

Physical Description
Date Built:
- 1900

Construction:
- Timber frame bank barn
- Vertical wood siding
- Metal gambrel roof with lightning rods
- Fieldstone foundation

Harry Christopher, Achie Helffrich, Fred Dohm, Les Gore,
Faye Dohm, Perry Christopher, Bert Kroupa

In the words of Frances Carroll McCaw

Frances Carroll McCaw reminisced with fond memories of living on the farm. In the 20's a tall tower was built just down the road on the Charles Carroll farm. It was called the Grand View Observation Tower, but was always known by the natives as Frederick Tower. "There was a pipe at the top of the tower running down to the ground where you could donate coins for the privilege of using the tower. The Carrolls would collect the money at the end of the day. We used to gather at the tower, and it became a social gathering place for young folks. There was a concession stand at the bottom of the tower. For 5 cents you could get pop, ice cream, or candy. Many downstate people came to climb the steps to the top of the tower. It was the place to hang out after a long day of picking cherries. I was a bit of a rascal, and I remember climbing to the top of the tower one time and drinking a coke and taking 2 aspirins. It was supposed to get you drunk. But I waited…and nothing happened. It didn't work!"

The tower was dismantled in the late 1950's. The tower had the significance of being chosen as the official site for the celebration of the Blessing of the Blossoms which led to the start of the National Cherry Festival, beginning in 1923.

During and shortly after World War II, the tower was used by volunteers to watch for enemy aircraft and was one of such observation sites nationwide. Most of the volunteers were women, who took turns working six hour shifts. Mary Johnson remembers manning the post, which was equipped with binoculars to spot planes and a telephone by which to report the information to the Civil Defense Office in Traverse City. She remembers reporting just one airplane in all her time up in the tower. When the tower was no longer being used and was in need of much repair, Bernie Carroll, whose property it was on, disassembled the tower but left the cement platform which can still be seen in the Carroll front yard near the road.

The view from this farm is spectacular. Both bays can be seen. Directly in front of the Lyon farm a tower was built in 1922, however that was on the Carroll property. This property runs along Center Road on a ridge of land the highest on the Peninsula. A prominent businessman in Traverse City, a gentleman named Albert Frederick, ran a shoe store. According to an article in the *Preview* Newspaper, March 14, 1983, "In 1922, he (Mr. Frederick) decided an observation tower placed on some particularly scenic spot on Old Mission area would be a boon to the tourism which was just beginning to be a major business in the area. He was willing to provide most of the financing himself, and several other businessmen contributed small amounts. It was built in 1923, and called the Grand View Observation Tower. That was changed to Friederich Tower later in the year. The tower had about 200 tourists per day."

DAVE EDMONDSON BARN

Three strikes and you're out! This barn is the third one on this site. The two previous barns burned when struck by lightning. Lightning rods were added after the second barn burning. This large 36 x 56 timber frame barn was built in 1900 and is built on grade, meaning it has no basement or lower level. The beams on both sides of the drive floor are hand hewn and run the width of the barn, making them about 36 feet long. The barn has a gable pitch roof which is now sagging in the middle. It is used for storage of equipment and odds and ends.

Before Dave Edmondson purchased the farm in 1986, it was owned by Paul Johnson from Empire, MI. Dave, a fourth generation farmer, grows cherries, apricots, peaches, nectarines, and apples on the acreage. The development rights to part of his land have been sold to the local PDR (purchase of development rights) program. Dave maintains 40 acres on this farm.

Physical Description
Date Built:
- 1900

Construction:
- Timber frame barn on grade
- Vertical wood siding
- Asphalt gable roof
- Rock and cement foundation

This farm was known for many years as William Carroll's farm. Carroll's daughter, Frances McCaw, remembers that during the Depression times were tough and that her dad would not accept welfare, so every Saturday he would go to town to sell butter and eggs. He made enough money to buy flour, sugar, and coffee. They always had 2 cows and 2 pigs, so they ate a lot of pork, and had chicken for every Sunday dinner. The farm grew mainly cherries but also some peaches, apples, and prunes. Cherry pickers often lived in the barn during the season and came back every year. Frances said the workers became great friends with the Carroll family.

"We had a few bad times on the farm," Frances remembers. "Once the pigs fell into the cistern that was beside the barn but we finally rescued them. And one very dry season my dad fed wild cherry branches to the cows because the grass had dried up. He found them all dead the next day, because wild cherry is poisonous to cows." Frances said her brother and sisters were expected to help on the farm. They helped in the garden, picked cherries, and churned the butter. It was a BIG job to clean the churn and separator afterwards.

At the top of the hill on the Edmondson's farm sits a house with a spectacular view of both bays…east and west. Dave had the house hauled up there from Neah-ta-wanta in 1991. It was moved five miles all in one piece. He showed me an article that was in the paper and a picture of it going down Center Road. It was titled, "House on the move." It was an undertaking on a grand scale. The cost was $14,000 to move it, and it has now been completely renovated on the inside.

Information provided by Dave Edmondson, Frances Carroll McCaw, and Mary Johnson.

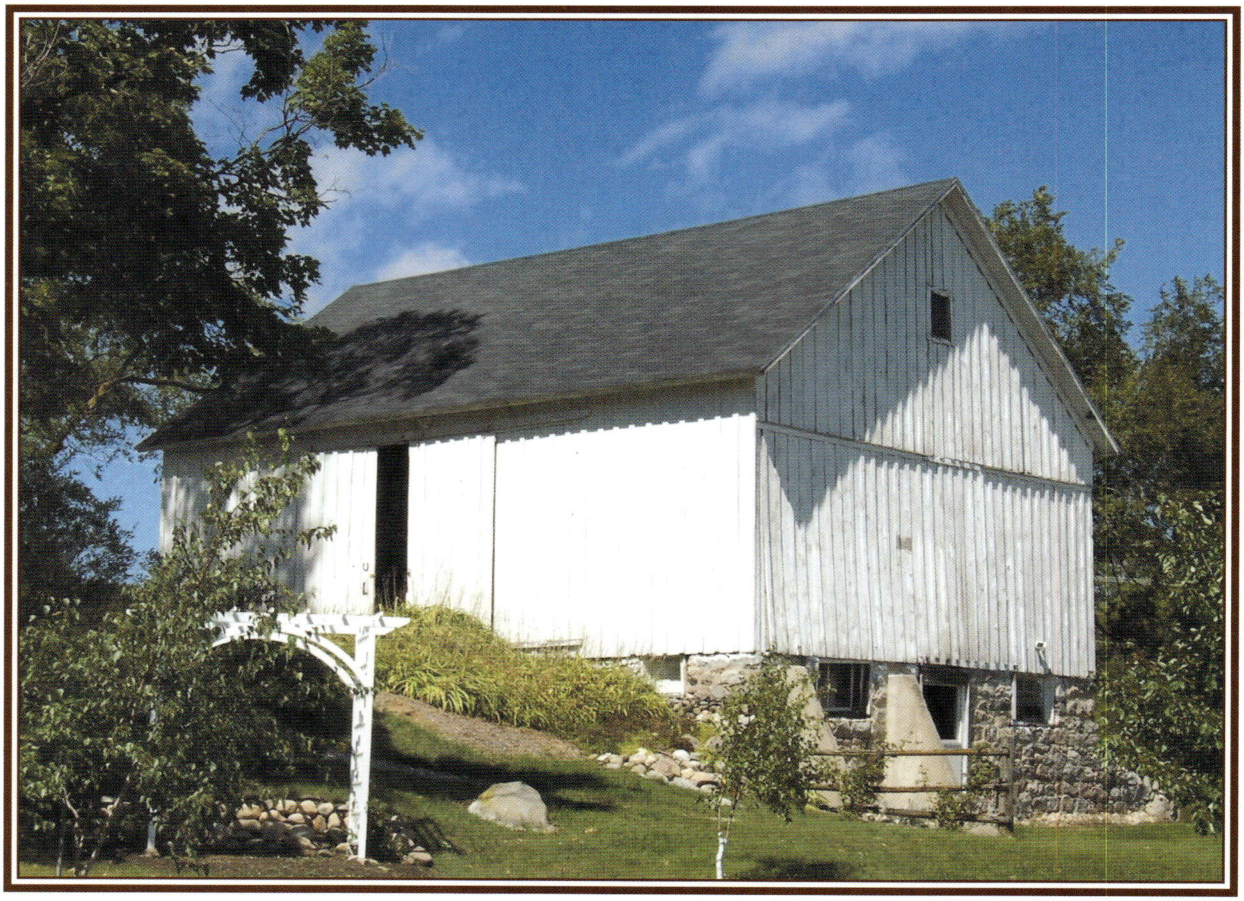

FLORENCE BARN

This barn sits up on high ground overlooking the scenic view of East Bay. The current owners, Ben and Janet Florence, have 11 acres consisting of a lovely, large 1892 farm home with extensive lawn and cottage gardens. It has been called Orchard Acres for a long time as at one time it was a cherry and prune farm. This 38 x 40 foot bank barn has a gable roof. The foundation is mortar made of beach sand and rocks. The age is uncertain but the Florences are certain it was built around the turn of the century or before 1900, in keeping with the age of the house.

When "big" Art McManus came from Canada to live here, he bought 40 acres. His name is on the 1895 plat map and was still on the plat map in the 1930's. He was called Big Art because there was another Art McManus on the Peninsula. He was a first cousin to George McManus Jr.'s grandfather who was called Little Art McManus. According to George, he was a colorful character who liked to fight. He was known in the lumber camps of northern Michigan and the Upper Peninsula as the champion fighter of all the lumber jacks.

Physical Description

Date Built:
- Probably before 1900

Construction:
- Timber frame ramped bank barn
- Vertical board and batten wood siding
- Asphalt shingle gable roof
- Mortar made up of beach sand and rocks

Eva Marie Crandall was the daughter of Art and Letita McManus. She died in 1996 well into her 90's. She was born in the farmhouse and taught at Stoney Beach School right next door to the farm. She married Deronda "Chum" Crandall. He owned a small gas station/grocery store along US-31 South, in what was then the heart of farmland. Today, the intersection is known as Chum's Corners, in his honor. Max Crandall, Eva's brother-in-law, started Max's Service, a thriving appliance business still today. Joan Zupin, Eva and Chum's daughter, said she remembered picking cherries on Grandpa's farm and remembered the big barn. According to Bill Lee, a former owner of the property, Eva's daughter Joan brought her mother back to the farm when she was in her 90's. Bill and his wife Jeri had remodeled the house and painted and reshingled the barn. Eva was delighted to see it.

Wallace Van Cleve Sr. bought the 40 acre cherry farm in the 30's. The Van Cleve family was from Thompsonville, where they had other orchards. When Wallace Sr. passed away in 1945-6 (grandson Wallace III isn't quite sure), Frank Youker managed the farm for Mrs. Van Cleve from 1949-55. This was Millie Shea's father and, growing up, she remembers being friends with the Van Cleve children on the farm. All but ten acres were sold off in 1955 to the Grishaws whose property adjoined the Van Cleves. Wallace Jr. was the next owner of the farm. His business was the R.C. Warren Cherry Processing Plant on the Peninsula and from 1952-84 it was a brining operation. The Van Cleve family would come up from June to August to stay on the farm. Wallace III said his dad repaired the crumbling foundation on the barn. It sits horizontal to the road, and the stone and mortar foundation can be seen. He added low walls to strengthen the foundation.

Bill Lynch bought the house, barn, and 4-5 acres in 1981. Dr. Bill Lee, a local dentist, rented the property for two years before he bought it, renovating and repairing the house and barn. Bill and Jeri lived there another 8 years before deciding it was not a safe place to raise children with the busy Center Road just beyond the green lawn in front of the house.

I love to see this big white barn with a huge lighted grapevine wreath on the barn door each Christmas. It was made from grapevines from the Florences' property.

Information provided by Ben and Janet Florence, Millie Shea, Joan Zupin, George McManus Jr., Dr. Bill Lee, and Wallace Van Cleve III.

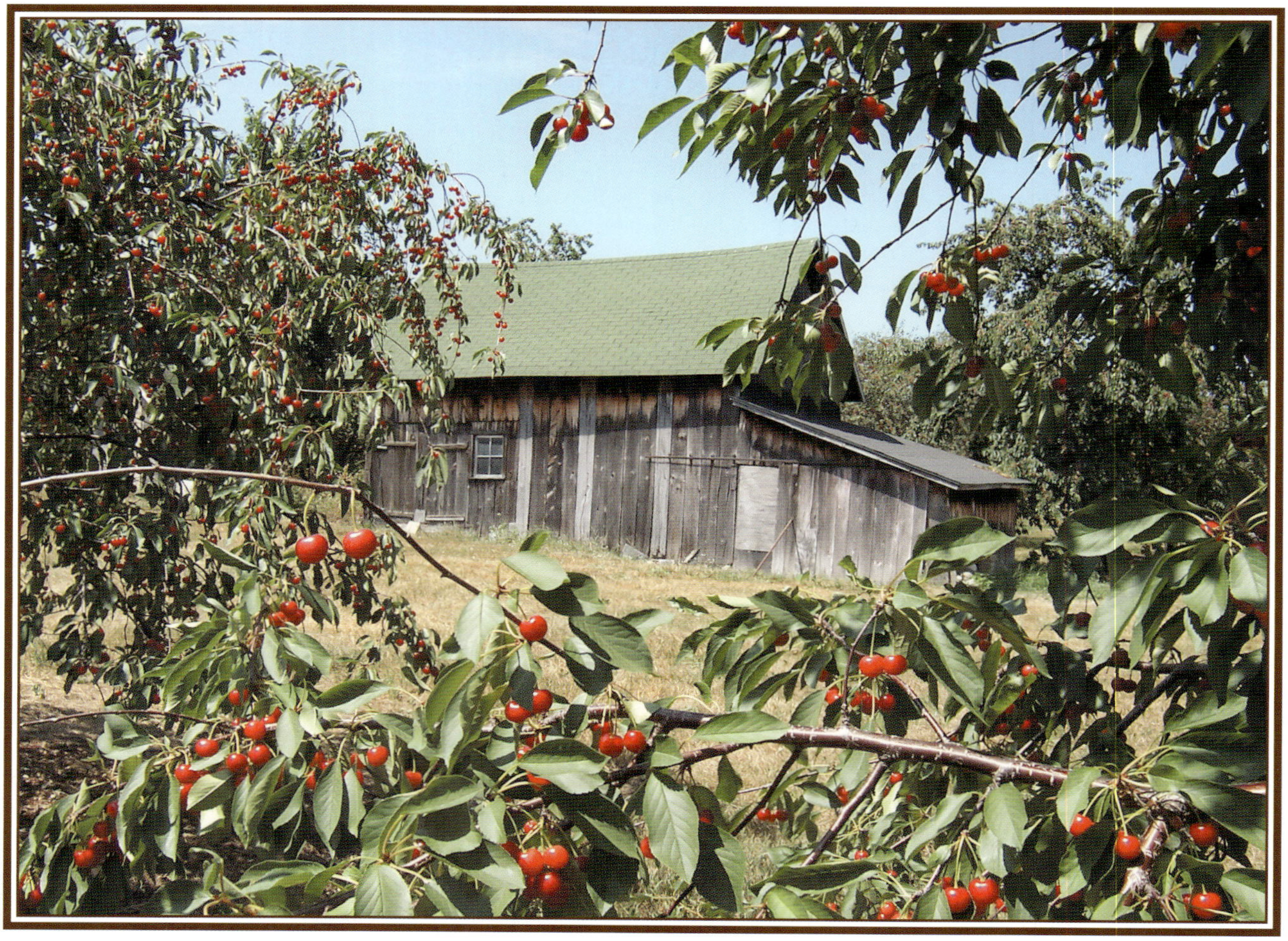

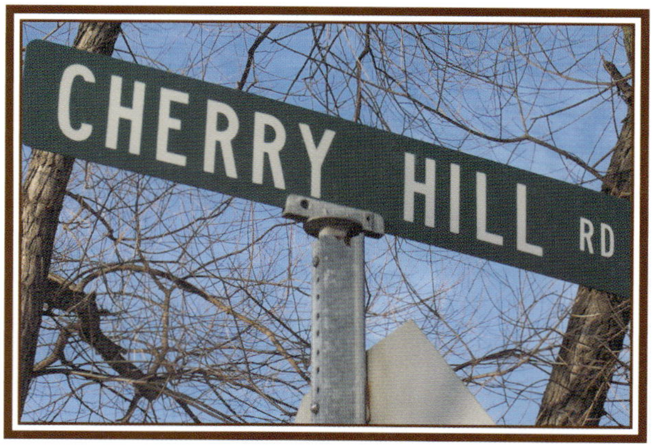
CHERRY HILL RD

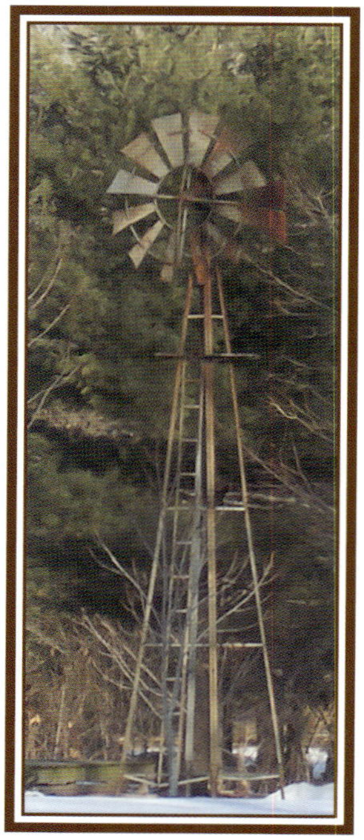

GOFF BARN

This humble little barn is all of 100 years old. It is set in the middle of an eight acre cherry orchard. When I asked Evelyn Goff why the barn was so small, she said that many of the barns and farms a century ago were small, 20 acres or less, as that was all one farmer could handle alone. Farm equipment that helped save labor and time was not yet available, and it was all the family could afford. These were self-sufficient farms, meant to make a life on but not necessarily to make a profit.

The first owner was Frank Christopher who owned many other acres along Center Road. Then the next owner was Carl Dell. The Goffs, who bought the property in 1958, also have a 25-acre cherry orchard south of Traverse City.

At one time there was a windmill on the property and a small, three-room house. The house and part of the barn were used to house Mexican-American families who came from Texas during cherry season to harvest the crop. The same families came back each year for eight years.

The small barn has a loft in the middle and a concrete floor on one side. The asphalt roof is about 10 years old. The barn has never been painted and is a weathered gray that stands nestled among a lovely cherry orchard. Today the barn stands vacant. The house was taken down a number of years ago, and the small windmill which was once on the property now stands in the yard of a home on Mission Road.

Physical Description
Date Built:
- c.a. 1905

Construction:
- Timber frame on grade with three side addition
- Vertical wood siding
- Asphalt gable roof
- Foundation: barn sits on rocks

The Goffs also owned another cherry farm, reportedly one of the best on the Peninsula, further down the Peninsula on Cherry Hill Road. The road was named by the Goffs' seventeen year old son, and the name is official now. The Goffs had 74 acres on this farm and lived there from 1978-1998.

Evelyn said that when they purchased their first cherry shaker, they hated telling their Texas friends they did not need them to pick anymore. She explained that the families would follow the crops, picking sugar beets in Minnesota, Cherries on Old Mission Peninsula and then down to Monroe to pick tomatoes. She said they liked to come to Northern Michigan because it was too hot in the summer in Texas!

Evelyn was a city girl who married a cherry grower, but she grew to love it. She laughingly told me, "There isn't a cherry farmer's wife that doesn't have an endless supply of cherry recipes to have you try. I think I have fixed cherries about every way possible."

Information provided by Evelyn Goff.

February 22, 1906
The project for building a canal from Torch Lake to Grand Traverse Bay is now underway and steps are being taken to secure the necessary consent of the property owners along the proposed route. By building the canal, boats could be run from the Bay directly to the villages on the shores of the lake.

34

JOHN HEMMING JR. BARN

John Hemming Jr. has a cherry stand next to the road during cherry season. He said he believes that the orchard in front of the barn likely has some of the most photographed cherry trees on the peninsula. It is picturesque in any season. Quoting from the *Record Eagle* in July 2005, John said a wedding party once stopped to snap pictures in his orchards. "A lady in a gown got up and snuggled real close to the trees," he said. John is a builder and a cherry grower. He said he picked up the construction craft after learning he could "swing a hammer" while he and a friend were building the cherry stand. Two Hemming families run the stand after picking the cherries off the trees that are nearby. This small 26 x 36 foot barn was built in 1900 and was used for storage of hay and also housed chickens, and, early on, other animals as well. It is in good shape and painted a nice bright red. Reportedly a small barn that was attached to the north of the present barn was the first building in Mapleton. John Hemming Sr. bought it in 1960.

This property of 28 acres was in the Lardie family for many, many years. I.F. Lardie shows ownership on the 1895 plat map. At one time the house on the property held the Citizens Telephone Exchange. However, the Citizens Telephone Company started at Old Mission at the Porter House, now called the Old Mission Inn. This one was run by the Porter family until 1926. The Citizens Telephone Company was an independent telephone company and the current owners of Old Mission Inn, the Jensens, have an old metal sign stating as such. Ike Lardie, I am assuming, started his Citizens Telephone Company some time after 1926.

However, there was another telephone company across the road, called the Mapleton Telephone Company, for a time, later becoming Peninsula Telephone Company. It cost 10 cents to make a call to anyone on the opposite phone company lines as it was long distance. Some folks had two telephones in their house…one for each company. The two telephone companies existed side by side until the 40's when Citizens was discontinued. In 1949 Jack and Vi Solomonson came to manage and operate the Peninsula Telephone Company in their home at 12450 Center Road, and have been the proprietors ever since. Jack Solomonson gave me a great deal of this information and said the switchboard for the telephone company was in their living room for years and they answered calls twenty-four hours a day.

The house on the Lardie property caught fire in the 50's, but the fire department was located right next door and managed to extinguish it before it totally burned. So, today what you see on the property is half a house. Ike Lardie was a photographer and had a large collection of photos, that went up in smoke as a result of the house fire. It is sad for many of the pictures would have shown us what life were like earlier in the twentieth century. Jack Solomonson said, "Ike must have had pictures of a million sunsets."

The small, neat house with the large history and a pretty little red barn is where John Hemming Jr. and his wife live.

Information provided by John Hemming Jr., John Hemming Sr., Angie Jensen, and Jack Solomonson.

> *Physical Description*
> Date Built:
> - 1900
>
> Construction:
> - *Timber frame on grade barn*
> - *Board and batten wood siding*
> - *Metal gable roof*
> - *No detectable foundation*

**1906 – Best buys of the week.
Women's Oriental Slippers, 25 cents at A.V. Fredrich
(the Tower on the Peninsula was named for him)
Fold up go-carts, $2.50 at J.W. Slater**

Holmes farms on both sides of Center Road in 1950's

MANNOR/HOLMES/LADD BARN

This barn stands alone in a field on Center Road, surrounded by tall grasses. It is no longer used but it once was a proud structure on a farm of 150 acres of fruit…cherries, peaches, and apples. It was used primarily for cattle into the 60's. The upper level held the hay. The foundation of a fieldstone silo is still there and old farm equipment can be seen on the main floor of the barn. This 33 x 44 foot timber frame barn has a gambrel wood-shingled roof and a large three-wall extension out back that is falling down. A newly remodeled farmhouse that once was part of the homestead is a short distance away from the barn. However the barn has a different owner who does not live in the area. No age was given on this barn. However, I am putting it in the "Turn of the Century Barn" category.

The original owner was Elisha P. Ladd who was a pioneer farmer, educator, and public servant (from *Grand Traverse Legends*, Vol. 1, p. 88). When he came to the Peninsula in 1853 with his wife Mary, they began staking out claims and clearing homesteads, along with other white settlers, even though the land had not been opened to them officially. E.P. Ladd was instrumental in getting the government to place the land on the open market. E.P was hired as the first school teacher in Peninsula Township when it was organized in 1853. He also served as supervisor of the township and county superintendent of schools along with being a farmer. He was held in high regard in the community when he passed away in 1898 at the age of 79 (taken from *Grand Traverse Legends* p. 88).

Son, Emmor O. Ladd, took over the farm where the barn now stands. E.O. graduated from Michigan Agricultural College (now MSU) in 1878. He taught school in various places for ten years, coming back to the Peninsula to became a successful farmer. He married Lizzie Stone in 1901. (His first wife died in 1898.) Lizzie, who also was a teacher, was the daughter of one of the original pioneers in Old Mission, William Stone. In a newspaper article in the *Grand Traverse Herald* in 1918, it stated that "he has been engaged successfully for many years in agriculture and became a leader in township affairs including township supervisor in 1900." In July 1918 he ran for State Representative for Northern Michigan, when he was 65 years old. He had the distinction of being the first legislator from the area and held the position from 1919 to 1924.

From Rebecca Tompkins Nothstine I learned that Emmor's wife, Lizzie, was grief stricken when E.O. died, and she would not talk to anyone. They had been an obviously sociable couple until his death. Rebecca told me that when she came to teach at the Old Mission school in 1938, she boarded at the Stone/Baldwin rooming house in Old Mission. Lizzie had gone back to her family home to live and refused to talk to anyone. However, she eventually talked to Rebecca. This seemed to end her long hiatus from communicating. In fact when Rebecca married Seth Tompkins in 1940, Lizzie was the Master of Ceremonies at the newlyweds' reception.

John Holmes married Lizzie's sister, Wilimina. The couple owned about 135 acres, according to the 1880 plat map, on the east side of Center Road. He started a family farm that remained in the Holmes family for over 100 years spanning four generations. The property sprawled on both sides of Center Road at times, with the different Holmes generations owning various parcels of land.

John's son Roy, and his wife Lula, bought the acreage on Center Road. Becky Wells, their granddaughter, remembers visiting the farm and reminisces about Grama Lula's beautiful hollyhock garden. She said one day Grama Holmes was in the garden in her robe when a snake slithered right between her legs. She lifted her long robe to her knees and commenced running down the drive and screaming, "Roy! Roy! Roy!"

Physical Description
Date Built:
- Around 1900

Construction:
- Timber frame barn
- Vertical wood siding
- Wood shake gambrel roof
- Fieldstone foundation on barn and silo, no longer standing

Becky said the barn held cattle in the extension on the front of the barn. The barn itself held hay and farm implements. Roy passed away in 1958 and Lula lived 20 more years on the farm.

Russell, son of Roy and Lula, bought the farm in 1944. While he farmed land on both sides of Center Road, he and his wife Katherine lived in a second house on the homestead, a stone house. In 1960, they moved to town…Old Mission, that is. They tore down the large, old Stone/Baldwin house on Mission Road and built a new "modern" house. Russell died in 1995.

Mannor/Holmes barn - 2005

MANNOR/HOLMES/LADD BARN

In 1975, Jack, son of Russell, and wife Carol, built a house in Old Mission, but farmed for 40 years on the west side of Center Road, where the old barn still stands. At one time they also lived in the old stately Ladd house on the property. In 1981 Jack bought his father Russell's farm across the road. Jack sold all but 30 acres of the property to Bernie Kroupa in 1995 and built a home on the Peninsula Drive. Jack passed away in 2004.

The Mannor family now own the forlorn looking barn and 15 acres, standing next to the lovely reconstructed farmhouse which once was a proud home and farm. The barn is no longer needed or useful. There is a sign over the barn that reads POVERTY HEIGHTS. It was placed there by Russell Holmes who was a jokester, according to Becky Wells. He was poking fun at the uncertainty of farming, probably expressing what many other area farmers felt at times.

The aerial picture, taken around 1945, shows Holmes farms on both sides of Center Road. The beautiful maple trees were nearly eighty years old, having been planted in 1876 to commemorate America's Centennial. Unfortunately all that remains of this homestead are pictures. Orchards have replaced the barn and houses. Most of the old maple trees have been cut down, too, as they had decayed or died.

Information provided by Rebecca Tompkins-Nothstine, Jack Solomonson, Carol Holmes, Bern Kroupa, and Becky Wells.

Typical orchard tractor with a front-end loader, most likely a Case - 1950's

ELZER BARN

Before Arnold Elzer bought this property with its turn-of-the-century barn, it was owned by Tony Zoulek and later by C. Case. Arnold purchased the farm in 1962 and grew cherries and apples on the 90 acres. Arnold and son John farmed together until Arnold retired to his house on Center Road. John remembers the barn once housed milk cows, and that they took the cream into Traverse City every week until the cream market "dried up." Then the family sold calves for veal. The barn also held riding horses and hogs.

Cherries have always been the main crop. John remembers they had many migrant workers living on the farm during the cherry season. They camped out in tents in front of the house. In the middle 1960's, John said they bought a limb shaker and the need for migrant help greatly diminished.

This bank barn has a gambrel roof, and sadly, the asphalt shingles on it are almost gone. In 1964, a tornado hit the barn. It went right over the middle of the roof. The force of the wind blew the cupola off the barn, and they found it in a field nearby. The barn has never been repaired. It once was painted red, and has a fieldstone foundation on all four corners of the barn. The rest on the foundation is on cement blocks.

> ### Physical Description
> Date Built:
> - 1900
>
> Construction:
> - Timber frame double ramp barn
> - Vertical wood siding
> - Asphalt gambrel roof
> - Fieldstone and concrete raised foundation

Arnold Elzer talked about his history and how he came to be a farmer on the Peninsula. He was in the Coast Guard during World War II and played an active role in the Pacific, returning to the United States in 1946. He was sent to the air station in Traverse City working with helicopters, and here he met his wife Betty at the county fair. Arnold said they were happily married for 57 years before Betty passed away. He said they never argued. Arnold quipped, "Arguments and war are just alike….no one wins." Before Arnold retired from the Coast Guard in 1962, he had tours of duty in Newfoundland, the Arctic, and Alaska. He moved back to the Peninsula with his wife and took up farming, carrying on his family's tradition. Arnold learned his farming skills on his family farm in Ottawa, Illinois. The family was engaged in truck farming. They made hot beds which consisted of a 4-5 foot wide wooden frame, with a foot of dirt and horse manure inside the frame. Into this was planted seeds and a glass frame placed on top. Then they let nature take its course. They sold plants as well as produce on their farm.

Arnold's son, John now has 60 acres mostly in cherries.

Information obtained from Arnold and John Elzer.

Tornado damage on roof—1964

JACKSON BARN

This tidy little gambrel roof barn—built in 1900—has a flaired eave and a metal roof and is of timber frame construction. It sits on 5 acres of farmland, and the barn is used mainly for storage. I could find no history of its origin or use, but it most likely was used to keep animals, cows and horses, at one time. It was once part of the Gray and Co. cherry station. What drew me to the barn was the big round tank on the north side of the barn that sets it apart from most of the area barns. I am told by Mr. Jackson, who has owned the property since 2001, that the tank was once a railroad tank car used for water storage. It stands out in the middle of a field, in all its bright red glory, in direct contrast to the cherry trees and other foliage that are nearby.

Information provided by Ronald Jackson.

Physical Description
Date Built:
- 1900

Construction:
- Timber frame bank barn
- Vertical wood siding
- Ribbed metal gambrel roof with flared eaves
- Poured concrete foundation

McMANUS BARN

What a pleasant sight this nice red barn is, with its white board fence, as you come up the hill on Kroupa Road. Tom and Susie McManus have lived and farmed this 20 acres for 40 years. The gable roof barn, built in 1900, was moved from the Kilmurry property in 1941. It is a nicely maintained, utilitarian barn. The white board fence holds Susie's eleven feeder beef cattle. The cattle have their own covered shed on the end of the barn. Susie says raising the cattle is her hobby and she loves it. Tom is a cherry and apple grower. He said his father and grandfather made farming their lives, and he has too. Both Susie and Tom say that the farm is a quiet, comfortable place to be.

Before the McManuses bought the farm, it was owned by Harry Zang. Mr. Zang grew a variety of crops, called general farming. Tom did not think he started to grow cherries before leaving the farm in the 60's.

Information provided by Tom and Susie McManus.

Physical Description
Date Built:
- 1900

Construction:
- Timber frame bank barn
- Vertical wood siding
- Ribbed steel gable roof
- Fieldstone foundation

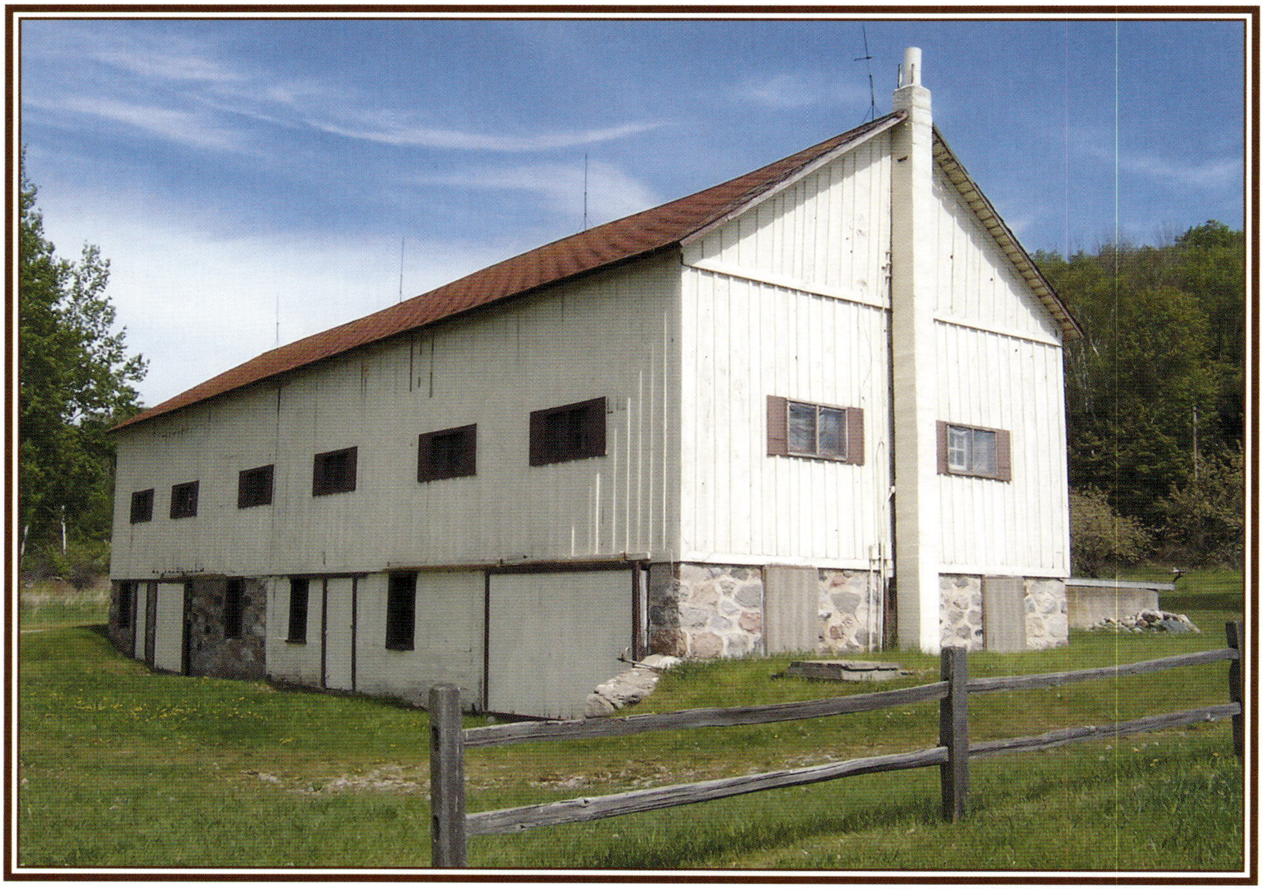

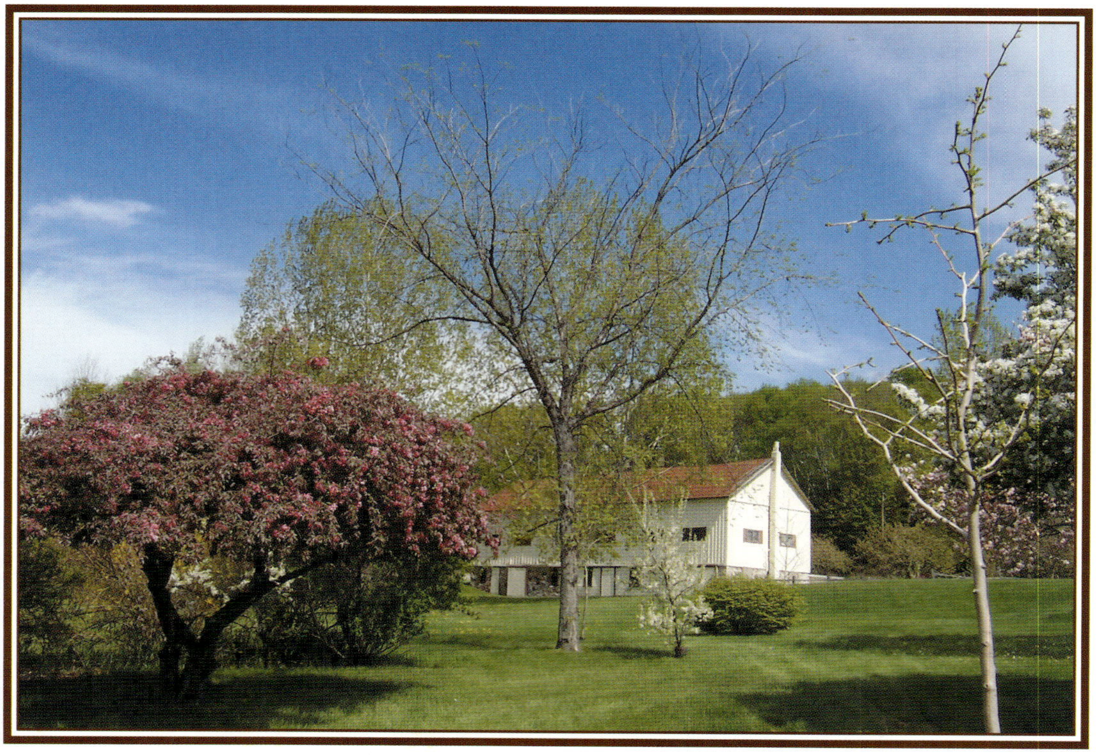

WILLIAM and SHIRLEY MILLER BARN

This beautiful big barn was built in 1900 by B.A. Benson. His brother was O.J. Benson, who came before him, and was known as a builder of barns and other structures in the area. The barn is 78 x 26 feet and, after much study and discussion, the Millers (current owners) and my husband believe it was actually two barns put together, both of the same vintage. Perhaps one half was moved there from a nearby location. At any rate, it is of timber frame construction from lumber taken off the 40 acre farm. There still is bark on some of the beams. The supports are hand hewn and the pole rafters are flattened on the roof side. The roof boards are of slab wood with bark showing on the edges. The first roof of wooden shingles shows through the widely separated boards. The second roof is of asphalt shingles. The roof boards were placed apart from each other on purpose to allow the wooden shingles to dry and let air get to them. The inside of the barn has been reinforced with cables and bracing and now the barn has been restored to very good condition.

There are three bays inside the barn on the main floor, with a loft on one side. The barn is a bank barn with the three levels showing on the front side. To enter the center bay you must go to the back of the barn. The cattle were kept on the lower level on the south side of the barn. It is white-washed inside and has a big door for the cattle to enter on the front (road) side. The gutters have been cemented in since cattle are no longer kept. There was a silo on the south end of the barn at one time.

Physical Description
Date Built:
 ◆ 1900
Construction:
 ◆ *Large timber frame bank barn*
 ◆ *Board and batten wood siding*
 ◆ *Asphalt shingle gable roof*
 ◆ *Fieldstone foundation*

The vertical board and batten siding is original and is painted a soft yellow, matching the lovely old farm house. The house has a beautiful fieldstone foundation, as does the barn. Shirley said the rocks were pushed and pulled out of the Bay which is approximately 100 yards away. Also, Shirley said, many of the rocks came from the fields on the farm, as the land was very rocky. She thinks the house has been redone at least two times. It is truly a lovely setting of 5 acres now, with lush green lawn and flowering trees around the barn and house.

The 40 acres that B.A. Benson had was fenced at one time, probably for his herd of cattle. He grew fruit because, when the Cosgrove family bought the farm from Mr. Benson in 1937, there was a good sized fruit farm in operation. They grew cherries, apples, plums, and pears. The harvested cherries went to the F&M Cannery in Traverse City. Much of the other fruit was sold to Gerber Baby Food Company.

Door secured with broken oar

Shirley remembers that, originally, the only water supply came from a spring on the property and was piped into the house and barn. But when migrants came to harvest the fruit, government regulations were such that the Cosgroves had to put down a well. They got their first electricity in 1946.

Shirley Cosgrove Miller has had an interesting life. During the school year her family lived in Lansing where her father was Superintendent of the Lansing Public Schools. When school was out in the spring, they all came north to the farm for the summer. She remembers working in the fruit...tilling around the fruit (called open cultivation), pitching hay, and many other jobs. When I asked what else her jobs entailed, she sighed and rolled her eyes. Enough said.

Her father retired to the farm in 1950 and introduced liquid fertilizer to the area. The company was called American Nitrogen. In 1984, William and Shirley bought the farm. Shirley said the barn is very important to her and she never plans to sell it. "It is part of my life!" she said. I can understand her strong feelings. The setting is beautiful and full of history.

All information provided by Shirley Miller.

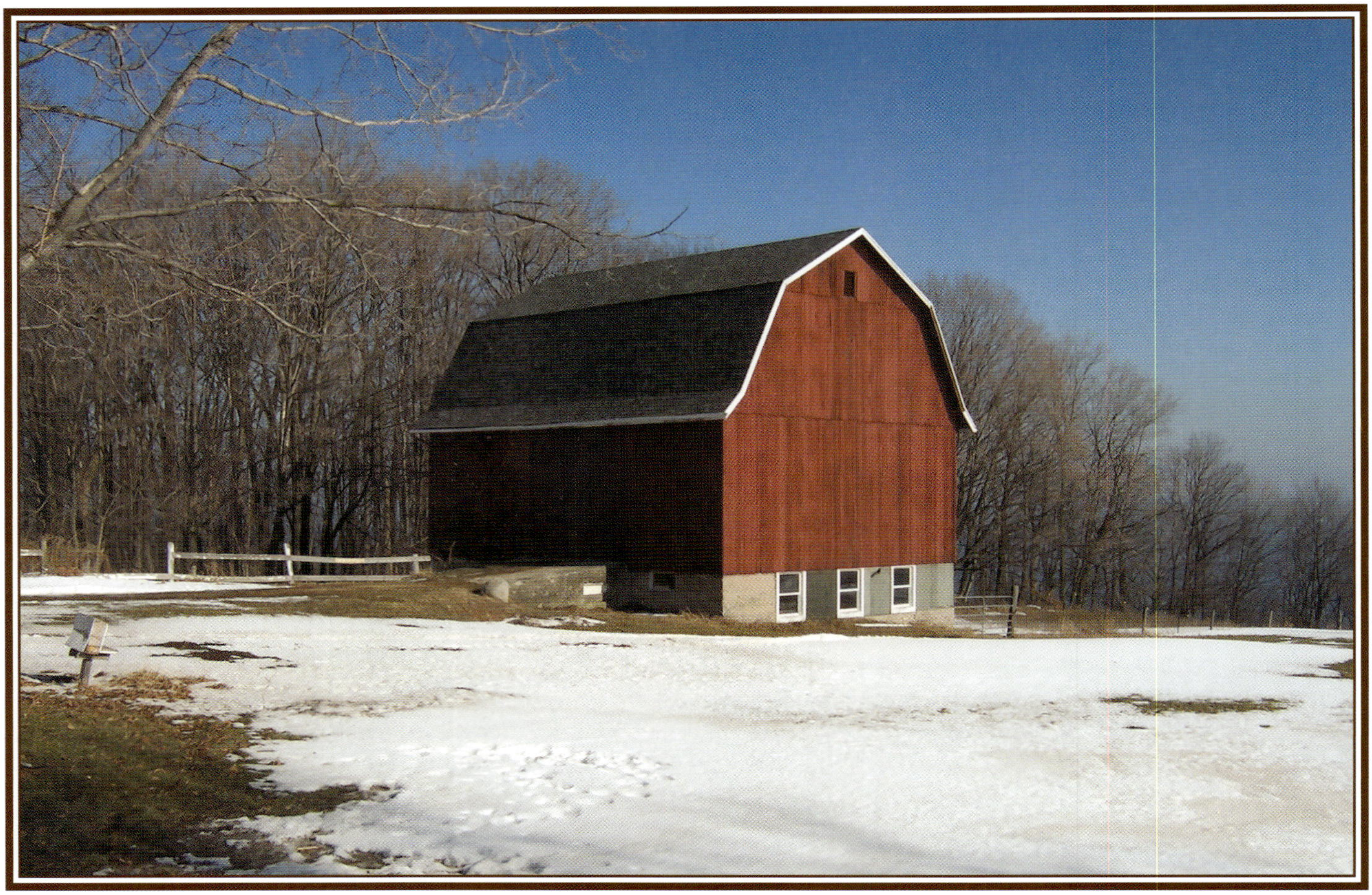

MISSION HILL BARN

This handsome barn sits high on a bluff overlooking East Bay and the remnants of a ship wreck. It sits near the end of Old Mission Peninsula. This timber frame barn is recorded as having been built in 1900. The asphalt shingled gambrel roof has a flaired eave and is in good condition. The barn has a concrete block foundation, with many stones and shells mixed in the mortar. This three bay barn is used for storage of hay on the main floor. The lower level has stalls with one of them housing a horse. The farm, along with other acreage in the immediate area, is still farmed, cherries being the main crop.

About the shipwreck, I gleaned this from an article written by Jane Louise Boursaw, August 20, 1998 for the *Record Eagle*. "Snowflakes swirled around the hull of the schooner Metropolis as she pulled out of Elk Rapids late one night on November 26, 1886, loaded with rough-hewn pine boards and pig-iron. Had Captain Duncan Corbett known that this blustery night would signal the proud sailing vessel's last voyage, he never would have attempted this one last run to Chicago before winter set in…the 124.7 foot schooner set out on her journey only to run aground near Old Mission Point around 3 a.m. Captain and crew managed to scramble ashore. Though there was a heavy northwest gale, an attempt was made to salvage the vessel…after two days they abandoned the effort, and the ship." The remnants of the ship, a few "ribs" nearly buried in the sand, is all that remains in the water today. It is still a favorite destination for boaters who like to search for the unfortunate ship remains in the clear blue East Bay waters. Up in the rafters of the barn, it is reported, there are still some boards taken from that ship wreck.

Physical Description
Date Built:
- 1900

Construction:
- Timber frame bank barn
- Vertical wood siding
- Asphalt shingle gambrel roof with flared eaves
- Concrete foundation with stones and shells embedded

Thomas T. Reese was a Welshman whose family owned the farm for a long time starting with the 1880 plat map. Every plat map after that showed the Reeses owning property at this location. In the 1930's, I. Reese's name appears, owning 45.5 acres. Thomas Reese married Alice Archer, whose father Joseph owned the property adjoining his property. The Reeses had three children, Ida, Martha Belle (called Belle), and Charles. Alice died in 1912, and Thomas in 1915, and Reese offspring lived on in the house and farmed the land—mainly cherries and raspberries. Daughter Belle married Robert Buell.

Charles Reese was a very enterprising fellow. He had a cider mill on the corner of Swaney and Mission Road for a time, and he worked for the Leffingwells property nearby, and on the Prescotts farm as well. The Reeses often had the help of hired men who lived in the house next door, where their mother had grown up. When Belle's husband Robert died in 1925, she remarried one of the hired hands. His name was William Welhusen, often called Whistling Willy. They were the talk of the town because Belle was 48 at the time and Willy was 21. But the marriage worked out, and Willy was with her, and farmed the property, until she died in 1968 at the age of 82.

Willy was well known as a bus driver for Old Mission Peninsula School for many years. Bill Welhusen was a happy fellow and well liked by the teachers, and especially loved by the kindergarten students who he delivered home each day. He always gave treats to the children on holidays, joked with them, and was a caring man, who made sure each child was delivered safely home, according to the former kindergarten teacher, Claudia Nowak. Norm Crampton remembers working with him when he drove a school bus for Old Mission School. They both had the kindergarten run. "Bill would pick up and deliver the kindergartners on the north end of the Peninsula and I headed south with my kindergartners." Norm said. Norm drove bus for 20 years and said that Bill was about 15 years older than him, guessing Bill was probably at least 65 years old when he retired. Bill remarried and moved to Missouri where his wife and he opened a resturant.

The farm now has a caretaker as the current owners do not live on the farm.

Information found in Elizabeth Potter's book, also from Jon Andrus, Claudia Nowak, and Norm Crampton.

Medical advice of a century ago:
The best time for bathing is in the morning, either before or after breakfast.
The vapor bath is an almost infallible cure for a cold.

The Rose farm bungalow

ROSE FARM

In a November 1905 article in the *Grand Traverse Herald* newspaper this was reported: "Rose Tompkins, daughter of one of the leading farmers on the Peninsula, Seth B. Tompkins, married W.E. Wilson. A very pretty wedding took place at Old Mission and Miss Tompkins was charmingly gowned in white silk. Reverend LeRoy Warren performed the wedding ceremony after which a wedding breakfast was served. The couple left on the morning train for a trip to Detroit and Chicago."

This 60 acre farm has since been known as the Rose farm. Rose's father gave her the property but her brother, Murry Tompkins, farmed it. The barn is a 30 x 40 foot gable roof bank barn built in 1900. It has board and batten siding and has been reroofed and maintained by the current owners, David and Sara Taft. The barn is no longer used except for storage, and the home on the property is rented.

Rose's husband Willis Wilson was the founder of Wilson's Furniture in Traverse City, which started in 1914, and was a thriving business in Traverse City for many decades. He was also a president of the Peoples State Bank, one of the first banks in Traverse City. Murry and wife Lulu lived on the farm as the Wilson family lived in town. The farm produced cherries, apples, peaches, plums, and prunes.

Physical Description
Date Built:
- 1900

Construction:
- Timber frame barn
- Board and batten wood siding
- Asphalt shingle gable roof
- Cement and stone foundation

The Wilsons had three daughters, Louise, Helen, and Beatrice. They became joint owners of the tenant farm as time went by. After Louise died, sisters Beatrice and Helen owned the farm. Beatrice married Frank Dewey Leonard and the couple had two children, Sara and Frank. As children in the 1950's, Frank and Sara both told me, they enjoyed coming to the farm in the summers. Sara remembers playing in the barn, which had a loft. Frank remembers a little white house near the barn that they called the bungalow, a gathering place for cousins and Aunt Helen too, where they occasionally spent summer nights. Frank said they helped pick cherries during the season.

Frank and Sara both remember their Great Aunt Mary Atherton who had raised their father Frank Leonard Sr. She ran the Neah-ta-wanta Inn which was a hotel and summer resort. She was a colorful character and had some well known guests who came to stay at her Inn on Bowers Harbor. Frank said his father remembers Henry Ford visiting, and it is reported that Al Capone was a sometime visitor also. Frank and Sara both have homes on Neah-ta-wanta.

The little house still stands just south of the barn. Beatrice eventually became the sole owner in the 1970's and it has since been purchased by her daughter Sara and her husband David Taft. The acreage is now farmed by Dave Edmondson and the house is rented.

Oral information provided by Frank Leonard, Sara Taft, and Rebecca Tompkins Nothstine.

**March 23, 1909
(Taken from the Township Board Minutes)
Moved and supported that we remit to John B. Boursaw one dollar out of the Dog Tax Fund as his dog had died.**

Summer

Winter

SHEA BARN

This lovely barn once was a taller structure but the roof was brought down 14 feet when the expansive upper level was no longer needed to store hay. The attractive cupolas on the roof of the barn were from the time when hay was stored in the top level to feed the dairy cattle and were used as the barn's vents. This 30 x 60 foot plank frame bank barn has a thick fieldstone foundation and the frame is reinforced with steel. It was built in 1900 and was converted into migrant housing when it became a cherry farm about 1947. In the past, when the McMullen family owned the farm, it stretched from West Bay to East Bay. The house on the property dates from 1875. The Walkers were the next owners before Don and Millie Shea bought the 31 acre farm in 1960.

One interesting note that Mrs. Shea told me was that the original owners, the McMullens, acquired the acreage (about 120 acres) from a land grant given by Abraham Lincoln. Through the years various McMullen names appear on the land parcels and the acreage varied, showing up on the plat maps. Probably family members were given pieces of the original 120 acres and other parcels of land were added.

Because there was no school nearby at the time the McMullens settled the land, Mrs. McMullen saw the need and solved the problem by having a school in the family home until other arrangements could be made. Later, Stoney Beach School was built just a short way down the road. According to *Sprague's History*, Albert P. Gray taught school at Stoney Beach from 1879-82, so it was an early school. (However other information said Stoney Beach school opened in 1883.)

Physical Description
Date Built:
- 1900

Construction:
- Plank frame bank barn
- Vertical wood on gambrel ends with many windows
- The rest of the barn is fieldstone
- Gambrel asphalt shingle roof with cupolas and dormers

The inside of the barn is very interesting as the migrant apartments are still in place even though they are no longer used. Don Shea said, "There were 7 apartments on the upper floor and 6 down in the lower level of the barn to house the migrants. The apartments came furnished with beds, a table, chairs, a refrigerator, etc. Toilets and showers were out back of the barn. Windows were added to give the apartments more light." Mr. Shea was the manager of the Peninsula Fruit Exchange at this time, and his wife, Millie, was in charge of taking care of the migrants and their families. Mrs. Shea said she was raised on a cherry farm but wasn't used to dealing with migrants. She told me she soon loved her job as the migrant workers were so delightful to work with. They became friends, often visiting when they were in the area. The Sheas' son Michael does the farming now.

This farm is situated on East Bay at Stoney Beach. It got its name, Mr. Shea told me, from all the stones, rocks, and boulders that were hauled off the land in the immediate area and dumped into the bay, making a point out into the bay at this spot. He said Sam Kelly told him of hauling a stone boat filled with rocks back and forth to the bay. When the Sheas bought the property in 1960, he said it was rare to see a powerboat on the water, since the water was not used for leisure. He said that lakefront property had very little value, and he would be surprised if you could have sold it for $5.00 a foot. My how times have changed!

Information provided by Don and Millie Shea.

The Teahen wellhouse

TEAHEN BARN

This large, lovely red gambrel roof barn is a busy horse barn. It is a 36 x 48 foot timber frame barn and was built around 1900. In the 1880's the John McMullen family owned this land from West Bay to East Bay, about 169 acres. In the 1930's, forty acres were split from the larger acreage and sold to M. Hayes. The Morgan family who were local cherry growers/processors also owned it for a time. Victor Friday owned the farm for fifteen years. He split up about 40 acres of the farm to create Horizon Hills Subdivision in the 1970's. Jim and Roberta Teahen purchased 45 acres in 1987.

The barn has a raised fieldstone foundation. Like so many barns, the cattle were housed in the lower level. The stanchions were in place until about 1980. When cherries took over, sweet cherries were sorted on the main floor and then packed. Many cherry pickers were employed and the hog house was turned into housing for them. It seemed the hog house was built into the side hill with three sides underground and lined with stone. Jim Teahen laughingly recalls that the workers fought to live in this abode as it was always cool inside. After a long day in the orchards in the hot summer sun, it was the coolest place on the farm. It was used for housing until the early 1980's.

Physical Description
Date Built:
- 1900

Construction:
- Large timber frame ramp barn
- Vertical wood siding
- Metal gambrel roof
- Raised fieldstone foundation

Victor Friday, the owner just before the Teahens bought the property, is credited, along with his family, with building the Friday cherry-trunk shaker, starting a real change in the way cherries were harvested and greatly diminishing the need for cherry picking by hand.

The Teahens, who use the barn for horses, boarding and breeding them, usually have about 30-40 horses at any given time. Daughter Liz runs the operation with her sister Rebecca helping out. The girls give horseback riding lessons and have a lively business. There are eight stalls in the barn. There is a big enclosed riding arena attached to the back of the barn, which also houses more horse stalls. The property has a lovely stone wellhouse and a small, 120-year-old house. When the house was being remodeled, Jim said they found newspaper in the walls dating from the late 1800's. (Newspapers were a popular form of insulation in the early years.)

Joy Fowler with her horse

Jim grew up on a big horse farm (1000 acres) near Brighton, Michigan. He attended the Leelanau School in his youth and graduated from Michigan State University. In 1967, Jim and wife Roberta resettled in Traverse City. After teaching physical education for 2 and $^1/_2$ years, he decided he preferred working in real estate and on the farm. His wife remains in education as Dean of Campus Studies at Ferris State University.

Information provided by Jim Teahan.

Old cultivator

"Spirit hole"

WILSON/VOGEL BARN

Peter Wilson had six sons when he came to the Peninsula in 1870 to establish a homestead and farm. One son, William, got the original farm on Wilson Road. William's sons were Willard and James. James went into the farming business with his father's brother Arthur, who bought a farm on Montague Road in 1898. This farm adjoins the original Wilson family farm. Arthur, who never married, farmed 30 acres with James from 1938 until his death in 1952.

The barn is unseen from the road as it is down below the bank on Montague Road. By the time Arthur bought the farm, the small house and barn were already in place, the barn having been built around 1900. They raised all kinds of fruit…strawberries, raspberries, peaches, apples, plums, and cherries, with cherries being their cash crop.

Physical Description
Date Built:
- 1900

Construction:
- Timber frame bank barn with three side addition
- Vertical wood siding with spirit hole at peak
- Metal gambrel roof
- Fieldstone and concrete foundation

James married, and after several years of three generations living in one small farm house, James moved with his wife and two girls to Traverse City. He commuted to the farm until Uncle Arthur's death, after which the family moved back to the farm, living in the old house while a new house was built.

While this barn is old, it is in good condition. It is a bank barn with a basement level where the livestock were kept. They always had a cow or two and a team of horses which were used until 1952. Hay was stored in the upper part of the barn. Being a bank barn, the upper level seems like the main floor from the east side of the barn. The barn has three haying doors and a corrugated metal roof. It also has a "spirit hole" up near the gable end. The foundation is field stone cemented over on one end and a concrete wall on the other. There is a cistern located outside on the northeast corner of the barn and a guttering system fed rain water from the barn roof into the cistern. There are battens on the north end to help seal the barn from the weather. At one time, Helen Vogel told me, Montague Road went up the valley on the west side of the barn. But by 1895 it was switched to its current location, leaving room for a house between the road and the bank barn.

Helen remembers picking fruit. She called it "drudgery." She picked cherry lugs full of strawberries and cherries in season. She had cherry fights with her sister. Helen liked the outdoors, but disliked the unsteady income of farming. Helen said there was a huge maple tree out in the orchard that was used for direction when working in the orchard. That was where fruit was always packed because it was nice and cool in the shade. Unfortunately, it blew down in 1988, but the spot is still used as an intersection in the orchards.

Helen left the farm for college and became a teacher, but she always came back to visit her parents. Her father lived on the farm until he died at age 91. He had a remarkable memory for the past and for details, but was very deaf and hard to communicate with. Following the death of her parents and her husband in the 1990's, Helen returned to her roots and her Old Mission Peninsula home. She shared the picture she painted of the friendly old barn that is hidden from the view of all who pass by on the road above.

Helen shared this quip with me: "A favorite joke among cherry farmers was if one of them won the lottery, he would keep on farming…until the money was all gone."

Oral information and painting of the barn from Helen Vogel.

Painted in 2005

THE ROSE RIDGE FARM BARN

The Rose Ridge Farm barn is the newest and most unique barn on Old Mission Peninsula – an authentic Norwegian timber frame barn. This barn is the beginning of Dr. Joanne Westphal's dream to a create nonprofit MSU landscape architecture alumni project with the possibility of creating a continuing education center and 'field station' for the study of natural and sustainable design. Dr. Westphal, who is a professor of landscape architecture at MSU, began this labor of love in 1998 when she purchased 83 acres of prime agricultural land on Old Mission Peninsula.

In 2004, Westphal's landscape architecture students from MSU went to Norway to study with Kare Harfindal of West Norway Cultural Academy (instructor of sustainable timber framing and roofing). He agreed to come to Michigan to teach Norwegian timber framing in the fall of that year. He greeted 13 students and, with wood already cut from the site, the project began. Kare had given a precise list of barn materials ahead of time. "Not a single nail was used in the frame; and all wood materials came directly from the 40 acre site," stated Joanne Westphal.

This barn is unusual for it's 'green roof". The sedum roof system is a modern alternative to the traditional sod roofs found in Norway. A rubber membrane is overlaid with rolls of nylon mesh. Soil material and fourteen different kinds of sedum are planted on the roof! A cupola, topped with an ornamental dragon, brings natural light into the building. Cypress siding was salvaged from Bond Pickle Company brine tanks from Oconto, WI giving this barn a distinctive rustic hue. The siding was completed in the board and batten style for a traditional Scandinavian look.

The sedum roof

Hand hewn support cut into a large beam

Information provided by Dr. Joanne Westphal.

The Norwegian barn - January 2007

54

THE BARNS OF OLD MISSION PENINSULA
1900-1920

Edgar DeVol, Robert DeVol's father and his cows

Montague family barn - Purchased in 1934

Bos barn renovation 2005

BOS/MONTAGUE BARN

Richard Montague told me his parents, Tom and Lucy Montague, bought the farm in 1934, across the road from where he and his wife, Phyllis, now live. Both the house and barn needed repair when they moved in. Richard said they worked very hard to clean out the house and fix it up. They also painted and reroofed the barn with wood shingles, which was most likely built in 1916. There was always something that needed fixing as the buildings were in a run down condition. The barn was partially constructed of hemlock, Richard said. Sometime later the barn was roofed with metal.

I talked to Richard Montague about his life on the farm and he told me he always worked for what he wanted. When he was six years old he picked up apples from the ground for 2 cents a bushel. When he was nine years old he drove the horse riding on the disk. A disk breaks up the ground after it has been initially plowed. He said, "I was tied to the seat so I wouldn't fall under the wheels." He told me he used to ride a cutter, which is a sleigh. In fact, his first ride in a cutter was when he was 1 month old, all bundled up for the winter trip. Much later Richard drove the cutter and horse down the road to Archie Hall to dances there…with live music. This very same horse was later sold to one of the musicians, because he was too frisky for Richard's grandmother to handle.

Physical Description
Date Built:
- 1916

Construction:
- Timber frame bank barn
- Vertical wood siding
- Metal gambrel roof
- Fieldstone foundation

In the winters Richard remembers cutting ice from the Bay with his father and grandfather. Richard farmed all his life, and said, "I had the best life in the world right here on the farm. I played ball, horseshoes, had entertainment at Archie Hall and had no need to go to town (Traverse City)." Richard still works a big garden in the summer, though he no longer owns the barn across the road from his house.

The barn, now owned by the Bos family, has been completely renovated on the outside. It has new cedar board and batten siding and is painted a vivid red. A very spacious house is being built on the property up the hill behind the barn. The new owners look down on this tidy little barn and see a piece of history.

Oral information given by Richard Montague.

57

DAFOE BARN

Headstones found in the Parmelee Cemetery
overlooking Bagley Lake

DAFOE BARN

George Parmalee owned just about all of the north end of Old Mission Peninsula (about 500 acres), with the exception of the state land where the Old Mission Lighthouse is. When the elder Mr. Parmalee died, William Bagley, who was married to Mr. Parmalee's daughter Harriet, settled the estate, and in payment received the Old Mission dock and 80 acres that was to become the Bagley farm. There is a pond right in the middle of the 80 acres known as Bagley Pond. "Mr. Parmelee was buried behind this farm on a part of his land which he had set aside as a cemetery, and his grave stone stands there still, surrounded by the deep peace of the forest." (*Story of Old Mission*, Potter, p. 71). The cemetery still exists and is up on a hill overlooking Bagley Pond. Harriet Parmalee Bagley died at the birth of her second child. In 1888 William Bagley married Emma Pratt. Eventually, the Bagley family consisted of four boys and two girls.

According to William Bagley's daughter Katherine Marshall, "My father was a very hard working person. We weren't raised on a farm, so it came rather hard for him. He also owned the dock (at Haserot Beach) and in navigation season, that was where he was, and the boys sort of had to take over on the farm. The boys always had chores to do. I'm afraid that I was kind of a lazy child. I was excused from most work when I was young. My brothers' jobs were milking the cows, feeding and watering the horses, and currying them. We had a well in the barn; and under the barn floor we had a root cellar, and my father raised beets…sugar beets, (called mangolds), and so on, to give to the cows to whet their appetites in the winter. Our farm had a very light, sandy soil. And the trees grew well, but they required a lot of fertilizer. Commercial fertilizer. My father was an agent for the fertilizer company and he always used a great deal of it." (This was taken from an interview James Brammer had with Katherine Marshall in 1976.) A lot was packed into a small barn. It housed cattle, horses, a carriage, and had a workshop. The barn is of plank frame construction and is on grade. It has an asphalt shingle roof and shiplap horizontal siding. No date could be found as to the age of the small barn, but it resembles other barns built around 1910-20.

Physical Description
Date Built:
- Around 1910-1912

Construction:
- Plank frame on grade
- Shiplap horizonal siding
- Asphalt shingle gable roof
- Cement and stone foundation

Money was always hard to come by and the Bagleys could not withstand the economic depression in the 1930's. The land stood idle for a period of time. In the early 40's the Andrus family bought the farm. Then, in 1997 the Dafoes purchased 15 acres of the farm which included the lovely house and barn. The house is beautifully restored and the barn is in good shape and has become a grand workshop for Mr. Dafoe. This property has only had three owners.

Most of the information was taken from an interview of Katherine Bagley Marshall by James Brammer.

June 22, 1905
W.D. Bagley of Old Mission passed through the city last Thursday on his way home from Elk Rapids. Mr. Bagley was in the city looking after material for the rebuilding of the Old Mission dock.

DAYTON BARN

According to the present owners, Jon and Amy Jo Dayton, this 30 x 40 foot barn was built several years before the farmhouse, as is often the case in earlier times. It was important to have housing for the cattle, hay, and other animals as they were the family's livelihood. It is a timber frame barn with a gable end roof. The owners told me that it cost over $6,000 to preserve the barn with a new metal roof about 10 years ago. It is the original Ed Mathison barn. Frank Mathison, son of Ed, was born in this farmhouse, and eventually bought a farm just south of his father's farm.

Ed Mathison was a dairy farmer and grew cherries. Frank told the Daytons, who bought the property in 1984, that the barn was actually raised at one time and the stone and concrete block basement was then built under it. The lean-tos were also added at both ends. Original dairy feed troughs and stalls are still in the lower level.

The Daytons removed a large lean-to at the south end that was too far gone to repair. Now the nicely painted red barn, and the farmhouse, are in excellent condition, and present a very attractive sight as you drive out the Peninsula. Mrs. Dayton's lovely gardens and the peaceful setting make it hard to believe that busy Center Road is just two miles from downtown Traverse City! The barn is used for storage now, and part of the house is a craft shop—Amy Jo's Folk Art Studio.

Information provided by the Daytons

Physical Description
Date Built:
- 1910

Construction:
- Timber frame with shed extension on north side
- Vertical wood siding
- Ribbed metal roof
- Fieldstone foundation

Picture courtesy of *Barns and Blessings*, Robin Grothe photographer

BILL and MONICA HOFFMAN BARN

Ed Boprey and Bob Seaberg were farmers on the Peninsula. But they also built a number of barns and were known for their flaired eave roofline. This was formerly Bob Seaberg's barn and thus it is a good example of this style of barn building. The main barn is of timber frame construction and was built in 1920. In 1940, a cement block addition was added on the side, with a metal roof. The main part of the barn has been reshingled and has recently been painted a bright red. The farm was 80 acres, with 70 acres being in cherries. The acreage on the west side of Center Road is now owned by Marc and Deb Santucci. It was formerly the Bob and Monica Seaberg farm. The Garland family were the original owners.

Tom Hoffman (Bill's father) and George Kelly (Bob Seaberg's son) remember cutting logs on the property. The logs were used for firewood, timber for the mill, and flooring. The barn itself has heavy beams that run the width of the barn, made from wood taken from the property. The barn was used to house cattle, chickens, and hogs. George remembers helping build the cement block addition to the barn. He said his job was to mix the concrete and he did a lot of it. The barn is in good repair, very neat and clean, and used for storage now.

George Kelly told me that in the 1960's about 60 Herefords were shipped in from Montana, and Angus cattle were brought from Elk Rapids, Michigan. Every summer the cattle were hauled north of Elk Rapids where there were 80 acres for the cattle to graze. This was done for about 8-10 years. They grew hay and hauled that too. It seemed the only profit was the manure. George laughingly claims he enjoyed spreading it. Duck manure was imported from Hemlock, Michigan, for fertilizing the cherry trees on the property. George Kelly lived on the farm in later childhood years. He worked for Gleason and Co. as a truck driver hauling cherries to the processing plant in Traverse City. However, his stepfather, Bob Seaberg, got sick and he had to go back to farming, something he had not planned to do. That is the way it was when you lived on a farm. "I read an article from a learned man who described the normal work year for peasants as running to some 600 hours. Now I think we were in a sense peasants, and 600 hours of work would hardly see us through two summer months. I think the work time for farm adults in my youth would be closer to 50 to 60 hours a week, slightly less in winter and more in the summer. Not much in the way of vacations." (From *The Land and Back*, by Curtis Stadtfeld, p.12)

Physical Description
Date Built:
- 1920

Construction:
- Timber frame with concrete and cement block addition
- Vertical wood siding on main barn
- Asphalt shingled gambrel roof with flaired eaves
- Concrete foundation

Bill now farms his second farm, that of his parents Tom and Irene Hoffman, just down the road to the north. Both Bill and Monica are avid hunters. Monica has been hunting seriously since she was seventeen years old, having followed her father to his deer blind for many years. She has bagged a good deal of game in her years of hunting. Bill hunts deer too, but also hunts bear, coyote, fox and other animals. Bill has gotten so many coyote over time that he has had a full length coyote coat made for Monica.

Information provided by George Kelly and Monica Hoffman.

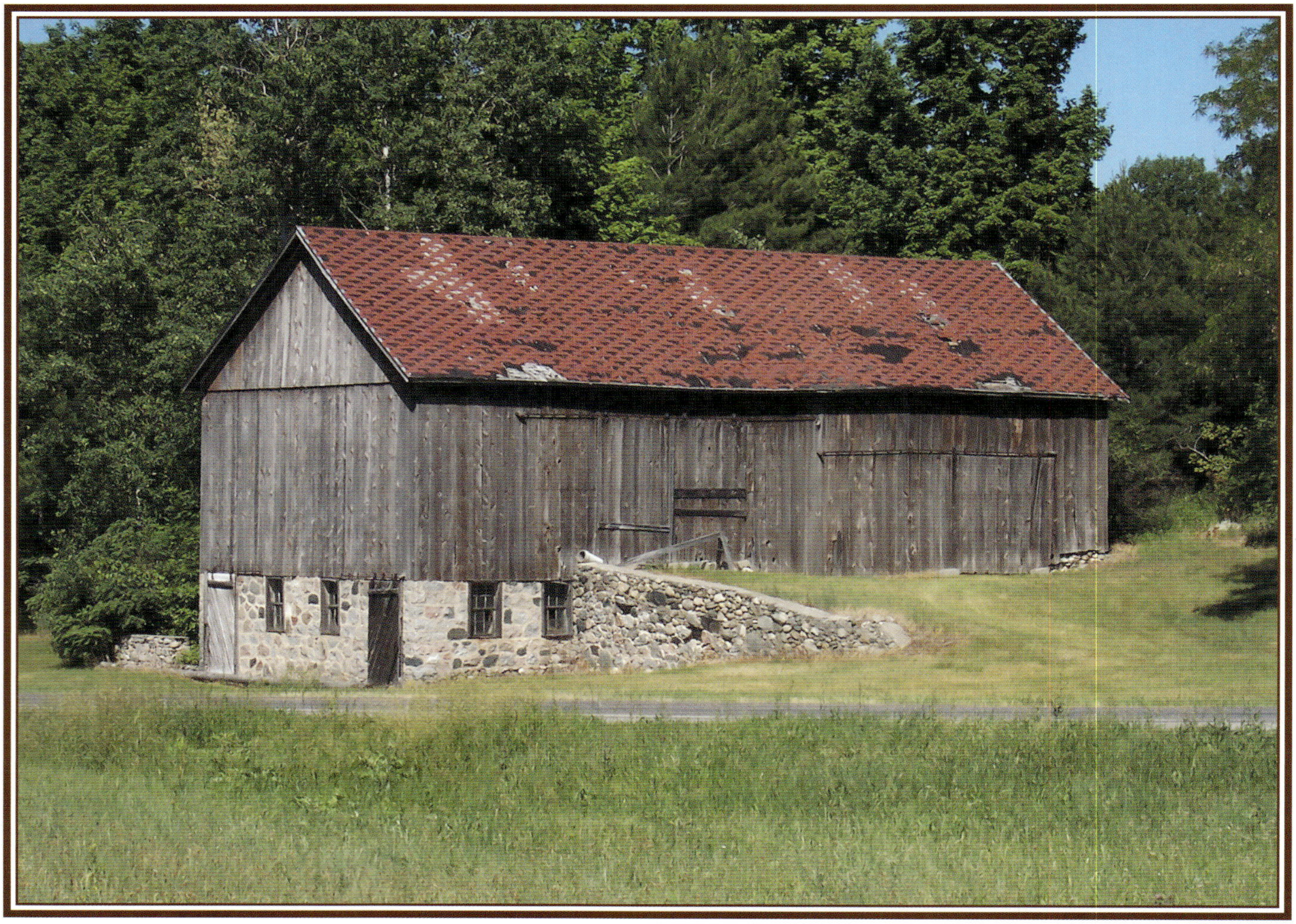

HARMON BARN

This scenic barn is perhaps the most photographed barn on the Peninsula, according to Bill Harmon. It sits close to the road and is in a lovely setting. The barn was built in 1906 and is a plank frame barn even though it gives you the impression of a timber frame when you see the laminated beams on the inside. Four 2x10 rough sawn planks are nailed together to make the main supporting beams in the barn. Bill pointed out a row of old licence plates dating from 1917 nailed on a wall inside which serve as battens between gaps in the planks. The barn has a 2 ½ foot thick fieldstone foundation with a cubicle cut into the stone on the inside. Here, a very old fire extingisher and other necessities are still in place. The 30 x 60 foot gable roof was reshingled in the 1950's. In 2006 Bill stabilized the inside of the barn and gave it a new red metal roof. It is now ready for another 100 years!

The lower level of the barn has homemade wooden stanchions for the 10-12 cows that were kept on the farm. The haying equipment, mow fork, etc. are still intact in the barn as is the mow door, which is the door that opens to drop hay down to the cattle below. There was a grainery in the southwest corner of the barn. Outside, at the back of the barn, you can still see a low stone wall where the manure was pitched through a barn door. This pine and cedar barn is banked on two sides and is in good condition. On the north side you can see a faint red patina of a barn that was once painted red. Today it is used for storing farm equipment. It remains close to Center Road as the road was widened a good deal over the years. The road came to the barn.

Physical Description
Date Built:
- 1906

Construction:
- *Plank frame barn with double ramp*
- *Vertical wood siding*
- *Asphalt shingled gable roof*
- *Fieldstone foundation*

One of the first owners was Walt Prussing who had dairy cattle and also farmed 60 acres, primarily fruit. Ozzie and Etta Herkner purchased the farm in the 1940's and added acreage, farming 200 acres on both sides of Center Road. They raised sweet and tart cherries and processing apples for Gerber Baby Food Company. Bill recalls that they sold about 2000 bushels to Gerbers for baby food. The rest of the apples were sold for the fresh market. The farm was a working farm called High Lake Orchards and was, by this time, being farmed by Bill and Judy (Herkner) Harmon. Then, in 1988 much of the acreage was purchased by Paul Nine and named Underwood Farms. A barn that was on the west side of the road was torn down to make way for the housing development.

The Harmons farmed the remaining 10 acres of property for 4 more years. Bill recalls that in the 1980's General Motors used the scenic barn in commercials for cars…Oldsmobiles in particular. After that, the hillside on the west side of the road was slowly inhabited with large homes and Underwood Acres subdivision development was born. The Harmons no longer farm but have retained their 10 acres on the east side of Center Road.

Judy Harmon was related closely to the Leighton family who lived farther out on the Peninsula. The story goes that her grandparents and many other families on the Peninsula would go across the ice to Elk Rapids in the winter for shopping. It was a shorter distance and easier to get there across the frozen Bay than going to Traverse City. Sometimes they used ice boats—heavy boats with a sail that skimmed across the ice.

Information provided by Bill Harmon.

July 1903
Around the area, while out of doors, the men are busy in haying and the women indoors are busy canning up raspberries and cherries. There isn't much time for social activities or trips into the city, but they know these chores will be finished in a few weeks and then they will go about with shopping and visiting.

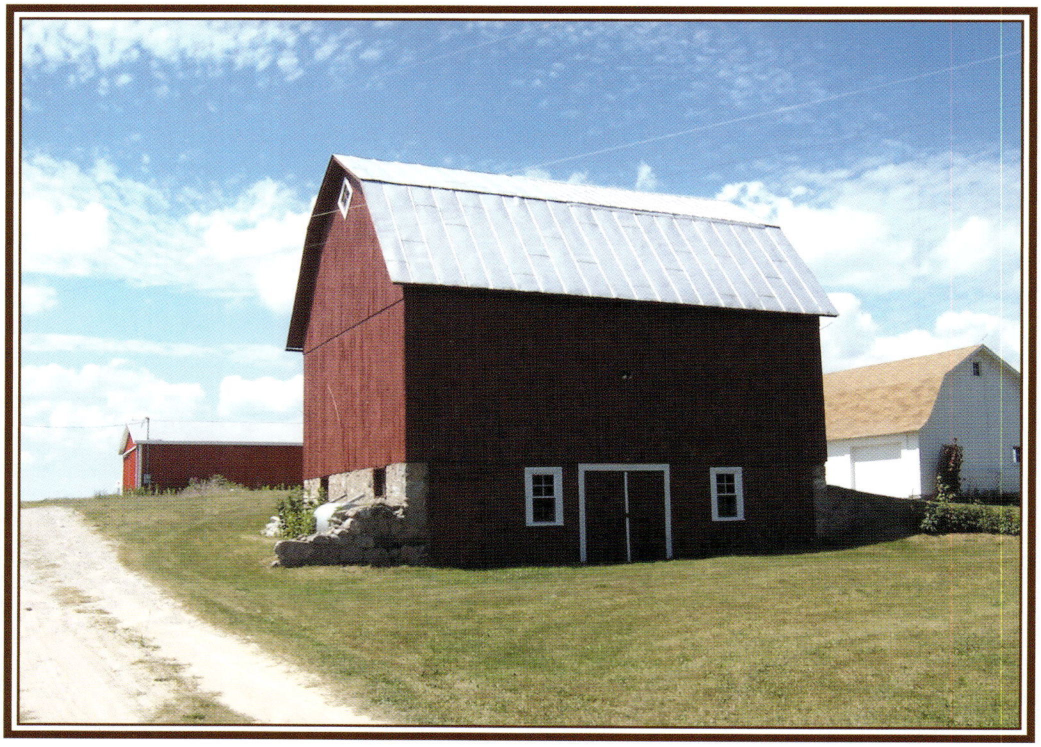

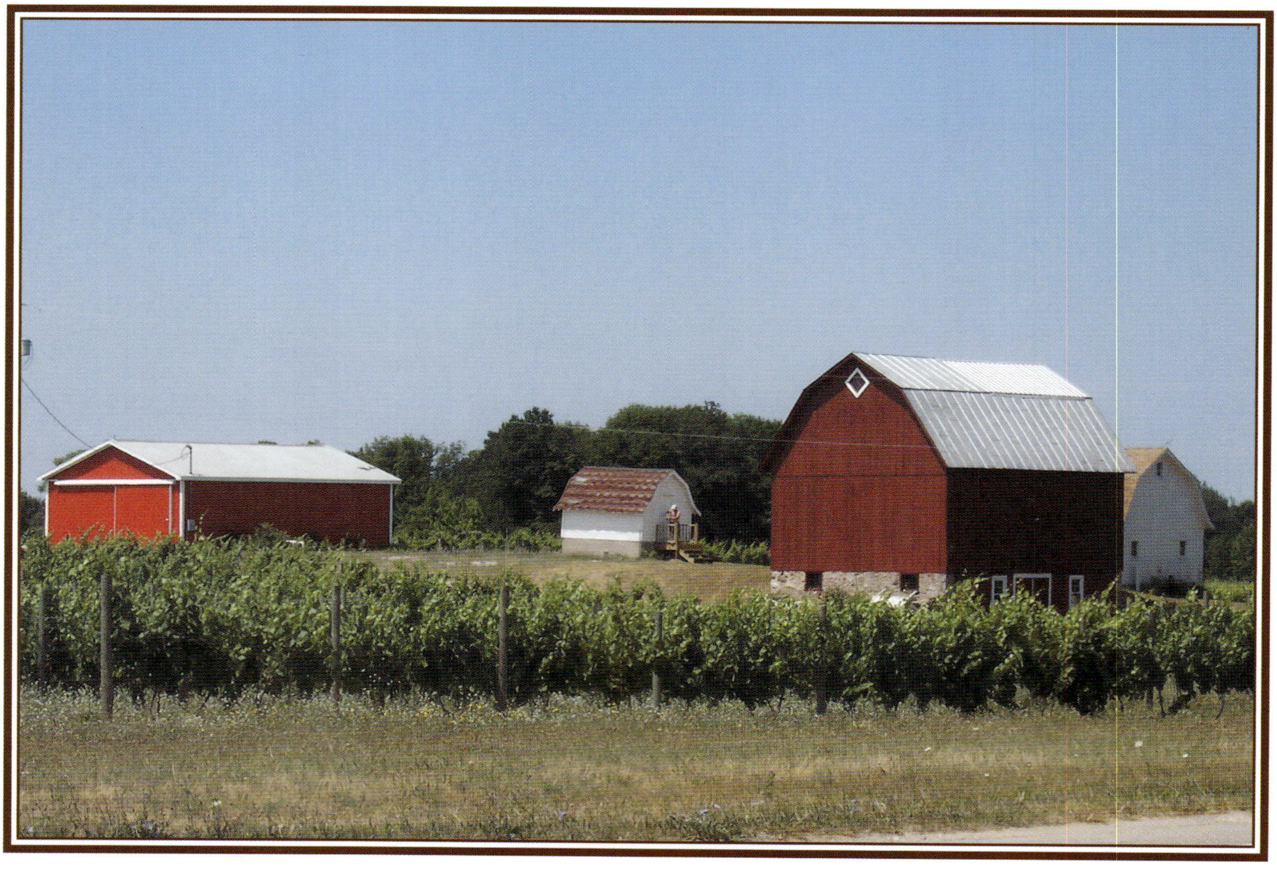

FIELD BARN

This barn was built in 1913 by Harry Christopher and continued in the family for four generations. In 1997 it was sold to Dennis and Sue Field. The farm itself encompassed forty acres on the west side of the road, with the house and barn and two 20-acre plots on the east side of the road…and just north of the present barn. Various family members have lived in these two locations. Basically, the farm has been a fruit farm, growing primarily cherries, but also apples, peaches, plums, and pears since the beginning of the fruit growing on the Peninsula.

The main barn is well taken care of and has two levels with a fieldstone foundation basement and a standing seam gambrel roof. There are 5 other buildings on the property besides the house. There is a chicken coop, a tack house, a pump house, the original one stall garage, and a second white barn with a flaired eave gambrel roof. The lower level of the attractive red gambrel barn originally housed cattle, and the stanchions are still in place. The upper level of the barn stored hay and farm implements. Hand hewn cedar joists support the barn floor. The farm has two cisterns to store water for washing cherries…one is under the tack house and the other is on the north side of the barn.

Physical Description
Date Built:
- 1913

Construction:
- Timber frame with cedar half log beams in lower level
- Vertical wood siding
- Standing seam gambrel roof
- Fieldstone foundation

Debbie Christopher Dunne grew up on the farm with her family. She remembers helping her parents who were full time farmers, as were their parents before them. The family picked cherries along with the migrant workers, and Debbie said they all worked hard. But they became good friends. She figures the family hired about 100 migrant workers and housed them in the barn, cabins, and other out buildings. Debbie and her brother and sisters drove the tractor, and did other chores on the farm. They were so busy in cherry season that they did not get to go to the Cherry Festival in Traverse City. However, in 1956, when she was in first grade at Old Mission School, she represented her school as the Cherry Princess, and Rob Manigold, was the Cherry Prince. That year the Cherry Festival was built into the farming schedule. She said this was the second year that Old Mission Peninsula School was opened…after all the other Peninsula schools were closed. The new school opened in 1955 and offered classes through the eighth grade.

The barn is used for storage now by the current owners, but Debbie said it was used for Halloween parties, for building "forts" and for rope swinging in the loft when they were growing up on the farm. She said that off season was not a time of rest on the farm. The family continued to work, as the trees needed pruning, new trees needed planting, and other chores had to be done.

Debbie remembers there was a windmill in the back yard that was not being used when she was growing up. One day a two-year-old cousin was found at the top, having climbed the ladder stairs on the side of the windmill. Her father tore it down right after the frightening incident.

Now the house is a Bed and Breakfast called Field of Dreams and the current owners Dennis and Sue Field have taken great care to keep the out buildings and the barn in excellent condition.

Information provided by
Debbie Christopher Dunne
and Sue Field.

I have found only two windmills intact on the peninsula.

Cherry Festival - 1956
Rob Manigold and Debbie Christopher - Prince and Princess

SWAFFER BARN

This old barn holds many fond memories for the Elmer Warren family, but it is not in good shape as the back roof has caved in and needs replacing badly. It was built in the early 1900's, probably 1910, according to Mrs. Elmer Warren. The original owner was William Gray, who had served in the Civil War and was granted the land by the government for his services. This medium sized timber frame barn is an on grade barn with a two shed extension, one on the eave side and one on the gable end. The siding is horizontal wood. The mid section has a loft on both sides. The roof, made of a corregated metal was added by the Warrens. The barn was lovingly taken care of when the Warrens owned it. Mrs. Warren said it was always white in color. Little by little over the years the acreage was shrunk from 40 acres to the 5 acres that the current owners, the Swaffers, own now. The back of the land is bordered by a subdivision of homes.

Mrs. Warren told me they lived on the farm for 32 years, spending most of their years in the large house on the property. The house was not always as large as it is today, but it was added onto as years went by. She said the house is over 95 years old. The Warren family acquired the farm from the Grays in 1946.

Physical Description
Date Built:
- 1910

Construction:
- Timber frame on grade barn
- Horizontal wood siding
- Corregated metal roof
- Foundation is poured concrete

Elmer Warren loved animals and had cattle, pigs, and other animals. He especially loved horses. He had Percheron draft horses from 1946 on, after returning from World War II. The Warrens' two daughters also had riding horses, and were actively involved in 4-H. Mrs. Warren said that one daughter recently wrote to her, reminiscing about her life on the farm, and what good memories they are. The barn, in this family's life, was a very important center. Unfortunately, Mr. Warren had a brain aneurism in 1969, and was disabled for 25 years. In 1978, the house and barn were sold to the Ware family who used it for boat storage.

In 1976, the house and barn and 7 and a half acres were sold to the Eldred family who used the barn for storage of boats, cars, and other vehicles, such as RVs. The barn was most important for Old Mission School during the cherry festival as the barn was used for building the float for the parades for many years. Marilyn Eldred said, "It was a fun time working together on the float and having picnics. It was a time of bringing the community together." The Eldreds owned The Bean Pot Restaurant for years, a familiar lunch place on the east end of downtown Traverse City. Richard Eldred utilized one end of the barn for his workshop. His hobby was making signs and doing other wood projects, including remodeling. In fact, when the Eldreds added to their house, Richard copied all the original molding of the century-old farm house in the new addition. The Eldred family lived there until the Swaffers bought it.

The Swaffers have owned the farm since 1992 and use the barn for storage. There still are stalls in the barn, from the days of the Warrens. The Swaffers have about 80 cherry trees. The large, lovely house has been redone and stands up a small rise from the old barn.

Fred Swaffer said his farming is largely done in Kaleva where he farms 80 acres organically with his partner, Chris Halpin. He has free range chickens, eggs, pigs, beef and goats. Fred was quoted in June 2005 in the *Record Eagle* as saying, "We all talk about open space and wanting to prevent urban sprawl, but the only way we can stop it is if farmers can make a living." This was at a picnic featuring local food, and in support of local farmers.

Information provided by Mrs. Elmer Warren, Marilyn Eldred, and Fred Swaffer.

Sign above small door

O'BRIEN BARN

This sturdy 40 x 30 foot gambrel roof barn was built in 1916 for William Dohm. It has a fieldstone foundation, a loft, and a lower level making it a three level barn. It was a general use barn at the time, housing cattle with stalls in the lower level and hay in the loft. Fred, his son, said his father bought the farm in 1908 to farm, but found it was hard to farm because of the hilly land. Fred said he just barely remembers the barn being built as he was four years old at the time. He does remember that in the winter the barn held the Peninsula snowplow. Perry Christopher and Archie Helfferich used it to plow the roads when they were working for the Peninsula Road Commission.

The Dohms sold the farm to the Willobys in 1920 and later it was sold to Perry Christopher. In 1999 Barry and Laura O'Brien bought the 3-plus acres from Dale Christopher. The barn came with the acreage and is used for storage by the O'Briens. The barn is well cared for as the O'Briens have spent a good deal of effort on their property, updating the house, inside and out, and a lovely little house out in back of their home.

Information provided by Barry O'Brien and Fred Dohm.

Physical Description
Date Built:
- 1916

Construction:
- Timber frame bank barn
- Vertical wood siding
- Flaired eave shingle gambrel roof
- Foundation is stone and cement

June 1905
Snow in June is not exactly in accordance with the previously conceived idea of the month of roses, but nature got ambitious Tuesday and worked in as many samples of weather as possible. The variety of showers was very satisfactory but in order to satisfy all, at 4 o'clock a little snowstorm started and lasted for about five minutes.

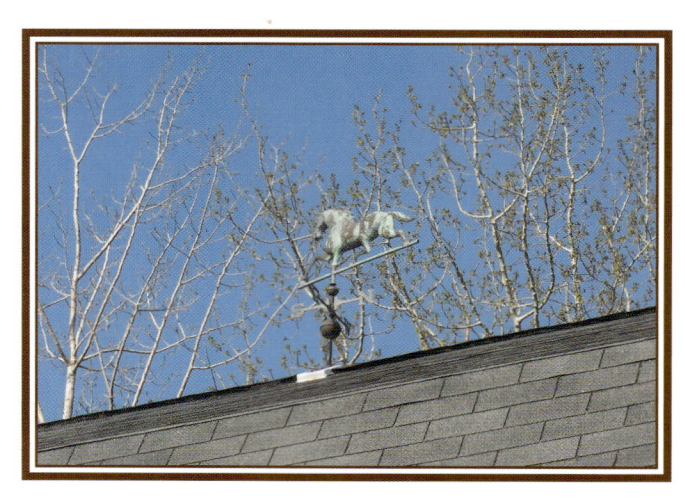

Horse weather-vane

DEAN and LAURA JOHNSON BARN

Oscar Nelson built this medium size 30 X 40 foot barn in 1909. It has a gambrel roofline and asphalt shingle roof. The barn was used to house cows and for general farming purposes. There appears to be a circular foundation of an old silo on the west side of the barn. Remains of what was probably an old chicken coop show under the lilac bushes on the northwest side of the barn.

Mr. Nelson married a widow, who came from Europe bringing her children with her. They had a child named Ruth. When Ruth married, her husband, Walter Rude, and she worked the 40-acre fruit farm which has the barn on it. They also raised and showed Tennessee Walker horses. They preferred the golden palominos, much like Roy Roger's horse Trigger. The Rudes put up wooden fences and remodeled the barn, adding five box stalls with generous sized mangers, dutch door, wooden floors, and nameplates over each stall door. One stall was reserved for the stallion and the other for mares. Many people who knew the Rudes remember a man with long, white whiskers named Asher who took care of the horses and took great pride in his work. He kept the barn in meticulous order. Ruth was an accountant at the Traverse City State Hospital. She was known to bring inmates home to help on the farm. Asher was one of them.

Physical Description
Date Built:
- 1909

Construction:
- Timber frame bank barn
- Vertical wood siding
- Asphalt gambrel roof
- Stone foundation

Through the horse related activities, the Rudes met the Brill family and maintained a close friendship throughout the years. The two women were instrumental in organizing horse shows at Bower's Harbor Park for a number of years. After Walter passed away, Ruth continued with the farm, but eventually she sold the horses. Ruth was not able to make much money on the farm through the years, but the last years she was there she had a block of Windsor Cherry trees that were in high demand. They proved to be very profitable for Ruth.

Dean Johnson had leased 35 acres from the Rudes for two or three years before buying the farm in 1993. It was said the farm was a bustling place, for along with the seasonal migrant help, the farm had trailers, which were either rented out or used for farm hands.

Dorothy Brill came to live and care for Ruth in her final years. When Ruth passed away in 1997, Dorothy was given five acres which included a ranch style house and the barn. She had many happy memories of the farm, and was delighted to have her neighbors use the barn for horses again, when Dean and Laura Johnson came. Later, Dean bought the 5 acres from Dorothy as well.

Today, the barn is well maintained but primarily used for hay and machinery. Perhaps it will hold horses again someday, as Laura is as much a lover of horses as Ruth was.

Information provided by Dean and Laura Johnson.

**1906 – Advice on deportment.
A gentleman never refuses to bow respectfully to his servants on the street, and a lady should do the same.**

1976

Bowers Harbor Park

11th Annual All Breed Horse Show

FRIDAY, SATURDAY AND SUNDAY
JUNE 11, 12, and 13, 1976 RAIN OR SHINE

Bowers Harbor Park is located 10 miles north of Traverse City on Old Mission Peninsula

Paid pre-entries must be received by May 29, 1976. No exceptions. Post entry fee is $2.00 added, per class.

Stall reservations must be in by May 29, 1976. No stalls will be reserved without payment in full.

Lehto Farm - 1912

Viola

Smokehouse

LEHTO BARN

This picturesque, medium sized gambrel roof barn sits down on Smokey Hollow Road in the midst of a mown field. It is well maintained and has a sturdy fieldstone foundation. It is 32 x 40 feet on two levels and was built in 1912. It has a ribbed steel roof, looks freshly painted, and has nice white trim. A fieldstone base of a silo still can be seen on the south side of the barn.

The property was originally owned by Mary Lucor, from 1869 to 1899. When George Jamieson bought the 40 acres it was to stay in the Jamieson family for the next 60 years. Floyd Jamieson, wife Mrytle, and their four children—Les, Ivan, Leon, and Viola—purchased the farm in 1901. Carl and Suzanne Lehto bought the property in 1971.

Daughter Viola wrote to the Lehtos in 1991 telling of living on the farm and of Grandma Myrtle giving the children milk and cookies as they walked home from the Ogdensburg school that was over the hill on Center Road. Viola MacAlary remembers, "The barn was built by our father, Floyd Jamieson, with some help from our uncle, Menton Willobee, who owned the Wells place up on the hill, and by a neighbor, George Helfferich. My dad bought a piece of timber land near Old Mission, and the lumber was cut and sawed by a then existing mill near there, and was hauled to the farm by horses. I can remember climbing up and walking the beams of the barn and getting a sound scolding."

Physical Description
Date Built:
- 1912

Construction:
- *Timber frame barn*
- *Vertical wood siding*
- *Ribbed steel gambrel roof*
- *Fieldstone foundation*

The barn held the horses and buggy in earlier times. Bruce Boursaw, a neighbor, remembers when there were about twenty head of cattle being kept in this barn. They were sharing the barn with the two horses, but the horses did not take kindly to the arrangement. To show their displeasure for the intruders, the horses kicked down the barn door, setting the cattle free. He said the Jamieson family had to chase the cattle all over the area.

While the barn housed cattle and hogs at one time in the lower level, hay was stored in the upper level. Later the barn was cleaned and used for migrant worker housing, as cherries were being grown on the acreage by then. The barn had four sets of double bunks and housed up to five families.

The old farm house burned in 1918 and the family was forced to live in the chicken coop while a new house was built. Taking no chances for a second fire, a fine square house of cement blocks was built in 1920. The blocks were made right on the farm in the barn.

One day when Carl Lehto, the present owner, was painting the side of the barn, a young couple stopped on the road and watched as he was up on a scaffolding painting away. Finally, the young man could stand it no longer and walked up to the barn and had

Ruins of a chicken coop that is now a beautiful sitting area

many questions about the barn. Finally, Carl asked him if he would like to help and, yes, the man climbed up on the scaffolding and commenced to paint, as happy as could be!

Other buildings on the five acre property are a Sauna (the Lehtos are Finnish), the old chicken coop ruins which have been made into a lovely garden sitting area, and a cement block smoke house. The house has been sided, it has a nice new roof on it, and a new addition. Suzanne Lehto has planted a lovely rock garden which is usually in full bloom all summer along the edge of the front yard.

Information and pictures provided by Carl and Suzanne Lehto.
Additional information by Cal Jamieson and Mr. Boursaw.

Lehto sauna

PANTER BARN

This charming 20 x 36 foot barn was built in 1918. It stands out because it is so brightly painted with a unique type of construction. It is a raised barn with a high foundation of glazed tile and is built right up to a steep hill. It was built by Clarence Samuelson, as was the house, in 1915. The farm was originally 40 acres—largely an orchard with some woods. Mr. Samuelson, who was a contractor, had a sawmill built on the property. The remnants of it are still visible in the woods. The barn is of plank construction with the upper part built with horizontal shiplap pine siding and is in excellent condition. The original cedar shingles are still in place on the barn. The gambrel roofline has an interesting turn up on the edges—called a flaired roof. The barn is not very large but has a lot to offer inside. The house is built of glazed tile also, but it is painted gray.

There is a stall for a horse in the lower level where the second owner, Dr. Pelto, kept his horse. He bought the property in 1963 from the Samuelson estate. The upper level has double doors on the back side where a hay wagon could drive right in. It is used for storage now and has a good sturdy main floor. Also in the upper part of the barn there are planks stored from Mr. Samuelson's sawmill, eighty years ago.

Physical Description
Date Built:
- 1918

Construction:
- Plank frame barn
- Horizontal shiplap siding
- Original cedar gambrel roof with flaired eaves
- Glazed tile foundation

Bob and Jo Ann Panter bought the property in 1973. Bob takes pride in caring for the barn, house, and other outbuildings, including a garage and a nice chicken coop. His property consists of three acres. Mr. Panter and his wife are former owners of Rainbow Book Store. In his previous career, Mr. Panter was a teacher at Old Mission school 1963-73 and the principal there from 1974 to 1987!

Farther up the hill behind the barn the land has been developed into subdivisions. Though the barn sits at the back of the Panters' property, it is a most pleasing sight as it is so unique and beautifully cared for.

Oral information by Robert Panter.

72

PELIZZARI BARN

As you drive up the bumpy road to this large, almost hidden barn, there is a lovely rock wall on the north side of the driveway. This 30 x40 foot barn has hewn timbers and was used to house cattle and chickens. It has a loft for hay and feed. It also has a 24 x 30 foot shed extension. Built in 1910, it has a gable roof with a faded red patina of an old, old barn. It has a metal roof and seems in good shape. It probably has many stories to tell if it could talk. During cherry season 50 pickers would come to harvest the crops, Mr. Eugene Pelizzari told me. They stayed in an outbuilding or in tents on the property. Eugene got their first cherry shaker in 1967, which eliminated the need for many of the pickers he previously used.

Eugene said they kept busy on the farm with gardening, harvesting cherries, and all the work with the livestock. The farm consisted of 62 acres in the 1920's when Eugene's father, A.I. Pelizzari, had the farm. In 1939, they had 40 acres in cherries, apples, and some hay. Then it became a farm of 40 acres of cherries as some of the land was sold off.

Eugene was in the Army for three years during World War II, 1942-45, in the European theater. Eugene said he took over the farm in 1951 and was done farming in 1990. His son John and grandson live on the farm now. The barn is used for storing sports cars and speed boats instead of tractors, trucks, and shakers. Time does change the "necessities" of life!

Information provided by Eugene and John Pelizzari.

Physical Description
Date Built:
- 1910

Construction:
- Timber frame bank barn with extension on south end
- Vertical wood siding
- Metal gable roof with sloping roof on shed
- Concrete foundation

RYCKMAN BARN

Marti Hyslop, a long time resident of Old Mission Peninsula, remembered this barn well. It was built in 1912 as a dairy barn for her father, Dr. Prescott. He was a dentist by profession practicing and living in Chicago, but the family would come to the Peninsula every summer. The family had purchased The Pines Resort in 1900 and ran it along with the farm. Hired hands had to do much of the work on the farm and at the resort. Marti remembered traveling back and forth to Chicago a good deal. The trip took 2½ days by steamer.

When Dr. Prescott retired from his practice, he became a farmer and built the house on Brinkman Road for the family. Marti said her father loved farming and remembered that he raised "fancy" pedigree cattle, chickens, and horses. The farm was 40 acres and the family grew cherries. Even though they had a nice barn, Marti liked to go to the Howes' barn out on The Ridge which was at the end of Ridgewood Road. In Marti's words, "They had a big, big barn that was lots of fun to play in with my friends. It was quite fascinating and elaborate, as well as very large." It unfortunately burned down, as did the large home that was near it.

Physical Description
Date Built:
- 1912

Construction:
- Timber frame barn
- Vertical wood siding (only half of the barn remains)
- Asphalt gable roof
- Foundation: unknown

Marti again, "The big boat, *The Puritan*, came in regularly in the fall to pick up fruit, and then it would go into Traverse City. You could ride into the city if you could get someone to go in and meet you. The boat would have to go around the point and into Traverse City. In my memory they packed apples in barrels. The community of Old Mission had a cooper shop where they made the barrels. They would load them on and, once in a while, a barrel would fall off and end up in the Bay. All along our shoreline there are duchess apple trees that must have grown from some of these apples that fell off. It is so pretty along there in blossom time."(From *Memories Hidden, Memories Found*, p. 138)

Marti spent her summers at The Pines along with many friends and relatives until her death in the winter of 2005 at the age of 94. The Pines is a treasure to behold in its original condition with its original furnishings dating back to the turn of the century. The barn and the house where Marti grew up have not fared so well. The barn is in ruins and the whole yard is completely overgrown with brush.

The farm, the barn, house, and acreage were sold in 1941 to the Ryckman family. Just the memories remain of the barn that housed "fancy cows."

Oral information provided by Marti Hyslop.

SANTUCCI/KROUPA BARN

This lovely old gable roof barn sits back from Center Road on property that was owned by Joseph and Elizabeth Kroupa. They had 60 acres which they purchased in three different sections. The house was built in 1904, and in 1905 the barn was constructed by O.J. Benson, a barn builder well known in the area. The barn was open to the weather on both north and south sides for a good long while until recently when it was sold to Louis Santucci and Anne Alvarez, who are restoring the barn. The barn has a good, sturdy, metal roof and lightning rods, complete with glass globes, still in place.

Early on, the Kroupas farmed with horses and kept cattle in the barn. Their son Henry took over the farm in 1945. He was a fruit grower all his life. Henry farmed alone, utilizing many migrants during cherry season. They were housed in a wooden building, a Quonset building, and tents.

Howard Wheelock married Henry Kroupa's daughter, Elizabeth, and helped farm when he could. Howard estimated they used about 100 pickers during the cherry season. Howard, who was raised on Ridgewood Farm at the north end of the Peninsula, had experience in farming as his father was Ridgewood Farm's manager for many years. Henry passed away several years ago and the house was left vacant. The barn at the time was owned by Arnold Elzer.

Louis Santucci and Anne Alvarez now have about five acres with the house and the barn. Both buildings are in the process of being completely remodeled and restored. Jeff Reinhardt is doing the restoration and I notice changes and progress on the make-over nearly every day as I drive by.

Reinhardt acknowledges his eclecticism. He spent time in the military and then as an emergency medical technician in Detroit during a particularly grim period in the city's history. When he'd seen enough, he fled to Leelanau County. That was 20 years ago. Reinhardt eventually landed in timber frame construction—commonly used in barns. "I got shuttled into barn repair because nobody else on the crew wanted to do it."

Before

Now Reinhardt stays with barn repair "out of a respect for the people who build them." (*Traverse Magazine*, "For the Love of Barns", Sept. 2001, p. 47)

Information provided by Howard Wheelock, Tom Hoffman, Arnold Elzer, and Anne Alvarez.

May 2006

Physical Description
Date Built:
- 1905

Construction:
- Timber frame bank barn
- Vertical wood siding
- Metal gable roof
- Fieldstone foundation

Summer fun

The barn in earlier years

ROSI FARM

About 1905, a small town lawyer named L.O. Vaught discovered this vast, beautiful woodland and waterfront spot and purchased it all from a Mary M. Walker. When Vaught returned to his home in Jacksonville, Illinois, he shared his enthusiasm for his new land acquisition with friends. Consequently, eight families from Illinois bought parcels of the land. They spent their summers living quietly and simply, first in tents, and later in rustic little cottages along the shore. There was no electric power and the only telephone was at the tenant farmer's house. The Illini group, as they were called, were a close knit group who lovingly called this area "The Strip."

The caretaker and tenant farmer of The Strip was Floyd Wilber. His son Addison wrote a book, *Memories*, about his life growing up on the Peninsula. In it he said, "My dad could do anything. He could build or repair any framework structure and he was an excellent orchardist and gardener. Dad could build the houses from the foundation to the roof including the plumbing, masonry and septic work. Some of his houses had beautiful stone fireplaces and porches. (Many of these houses can still be seen on The Strip.) Dad loved to grow things too. His garden was big enough to feed all the summer people and included many varieties of strawberries, blackberries, gooseberries, raspberries, currents, rhubarb, and asparagus to name a few."

Physical Description
Date Built:
- 1920

Construction:
- Plank frame on grade barn with three sided shed on one side
- Horizontal wood siding
- Asphalt shingled, gambrel roof
- Concrete foundation

The story about the arrival of the Wilber family to the area is interesting. From Addison's book, "After a long tiring trip we reached our destination. My father was to be the manager of the Illini Orchards Company which had 100 acres of orchards and a mile long strip of waterfront on West Bay. This was in March 1918. Mr. Marshall (a neighbor) told us that when we got our goods unloaded, we should come and have dinner at his house. He told us that while the house had been vacant, he had tapped the maple trees around the house to make maple syrup, and we would find a half dozen pails of sap in the house, which we could just dump out. After the goods were unloaded, dad built a fire in the stove to warm up the house. A little later, my brother Floyd went upstairs and noticed that in one room, the wall was warm and red. Dad rushed out to the woodpile and got an axe. He chopped into the wall. He found fire. The water to the house had not yet been turned on, but we did have those pails of maple sap, which we doused on the fire to keep it from spreading. This gave us time to open up a rainwater cistern and dip out water to finish the job."

Floyd Wilber built the barn in 1920. Jane Highsaw, a summer resident of "The Strip," tells about the barn. "There were two big gray and white horses in the barn, old Molly and Dan, and a cow. Siefkin's (one of the Strip summer people) had a pony too, named Blackie. Joan and I rode Blackie almost daily, bareback, along the road or up and down the front path. There was always one family of kittens in the barn or woodshed to be cuddled, chickens to be fed, horses grateful for a handful of oats. The hired man was usually filling the spray machine with water from the tank by the barn, or sharpening tools on the big grindstone by the shop. I remember the farm as such a lively center of activity. My task was to telephone the daily Lardie (grocery store) order from the Wilbers' wall phone." Jane's comments are found in the publication, *The Strip*, by Bonnie West, 1974. The Wilber family was greatly loved and the summer residents were sad to see them go in the mid-1940's. There were several other tenant farmers until the farmland was sold in 1970.

The barn itself still has the original 1920 horizontal siding. It houses many of the original tools and implements from Mr. Wilber's time. It is not a big barn as it is 24 x 28 with a 14 x 32 three-wall addition. It has a great deal of charm, sitting surrounded by woods. Dr. George E. Shambaugh bought the farm property in 1923 from Vaught. After he passed away in 1947, the property went to a succession of family members. The barn was used very little after 1950 and started to deteriorate. In 1994, Bob Rosi, grandson of George E. Shambaugh, and his wife, Penny, became the owners. Since 1994, the barn has undergone extensive repair work to preserve the structure and character, and is painted cherry red. The farmstead itself is now ten and a half acres, deep in the woods, and still has virgin timber on the property. The Rosis have built a lovely new farm home where the old house once stood. Many buildings remain, which makes one wonder what stories each of these could tell.

Oral information by Bob Rosi and Jane Highsaw.
Photos courtesy of Jane Highsaw.

The original house

Picture taken in the 1950's

What's left of the barn

1900
Lew Swaney replied, when asked why he built the barn before he built the house, "You can make money with a good barn."

SWANEY BARN

This farm has been in the Swaney family well over 100 years. While only half of the barn is standing now, it still has a story to tell. George Swaney was James and Harriet's (Hattie) second born son. Twenty acres of land were deeded to him when he turned 21. The deed states that this transpired in 1896. The land, being next to his older brother Jack's 20 acres, was formerly part of his parent's 80 acres. Lewis Swaney, James's third son, was also given 20 acres just south of George's 20 acres. He eventually sold it for what was believed, at the time, to be "worthless land on the bay" (on Smokey Hollow Road). "What could you grow there?" was the popular thought at the time. To this day there is a Louis Swaney Lane going down to the Bay. (Name was spelled differently.)

In 1901 George married Anna Belle Lardie and he built the barn the same year. The house was built in 1902 and was added on to several times as the family grew to six children. George and Anna's eldest son, Elmer, did not want to be a farmer and went to town (Traverse City) to work. Elmer married Lillian Magee and they had five children. Leslie, George's third son, lived on the family farm and cared for his mother Anna after his father died in 1937. Anna passed away in 1955.

George's brother, Jack, who never married, lived just to the north on the family farm with Ted Ayres, a Swaney relative. They farmed the combined farms growing, among other things, cherries. Leslie who never took to farming, was called the Peninsula electrician as he did a great deal of electrical work for those on the Peninsula. He lived on the farm until his death in 1984.

Physical Description
Date Built:
- 1901

Construction:
- Timber frame on grade with large slant roof extension on back
- Vertical wood siding
- Metal gable roof over sawn boards
- Foundation: wooden beams on stone piling

Elmer Swaney's youngest son Gerald and his wife Ann now own the property, buying it from the Leslie Swaney Estate in 1984. Gerald graduated from Michigan State University in 1965 and went to work for NASA's Manned Spacecraft Center in Houston, Texas, as an engineer in 1968. In 1980 Gerry and Ann brought their family back to Michigan and settled in the village of Old Mission. They built their new home in 1997 on the original George Swaney property. The 1902 family home was purchased by Ted and Lucile Bagley who had it moved across the road. They remodeled into a lovely home with a spectacular view of East Bay.

The old barn out back was falling into disrepair and part of it came down about 1990. Only half of it remains today. In its entirety it was a T shaped timber frame barn with a gable roof. The smaller tall section had a loft for hay and the horses were kept in this part. The other part, that no longer exists, was a large shed extension for the chickens and another horse. Standing beside the remaining part of the barn is an old spray rig of bygone days.

Gerry and Ann's property consists of about ten acres now. They continued to raise cherries and had a fruit stand alongside Center Road in the late 1980's. The old cherry trees were pulled out when they built the house in 1997. Ann remembers the family growing three acres of asparagus and selling it throughout a three-year period.

Gerry has fond memories of the farm of the past. He said when his family came to visit Grandma Swaney, he watched for the big maple tree along the side of Center Road. He knew they were getting close to Grandma's farm when he spotted the tree. The big maple tree is still there.

Information and pictures provided by Gerald and Ann Swaney; also Pat Wolf.

1904-05
Mr. and Mrs. George Swaney of Old Mission spent last Sunday with Mr. and Mrs. J. Lardie at Mapleton. A substantial lunch was served after which the friends enjoyed singing together many of the old favorite hymns.

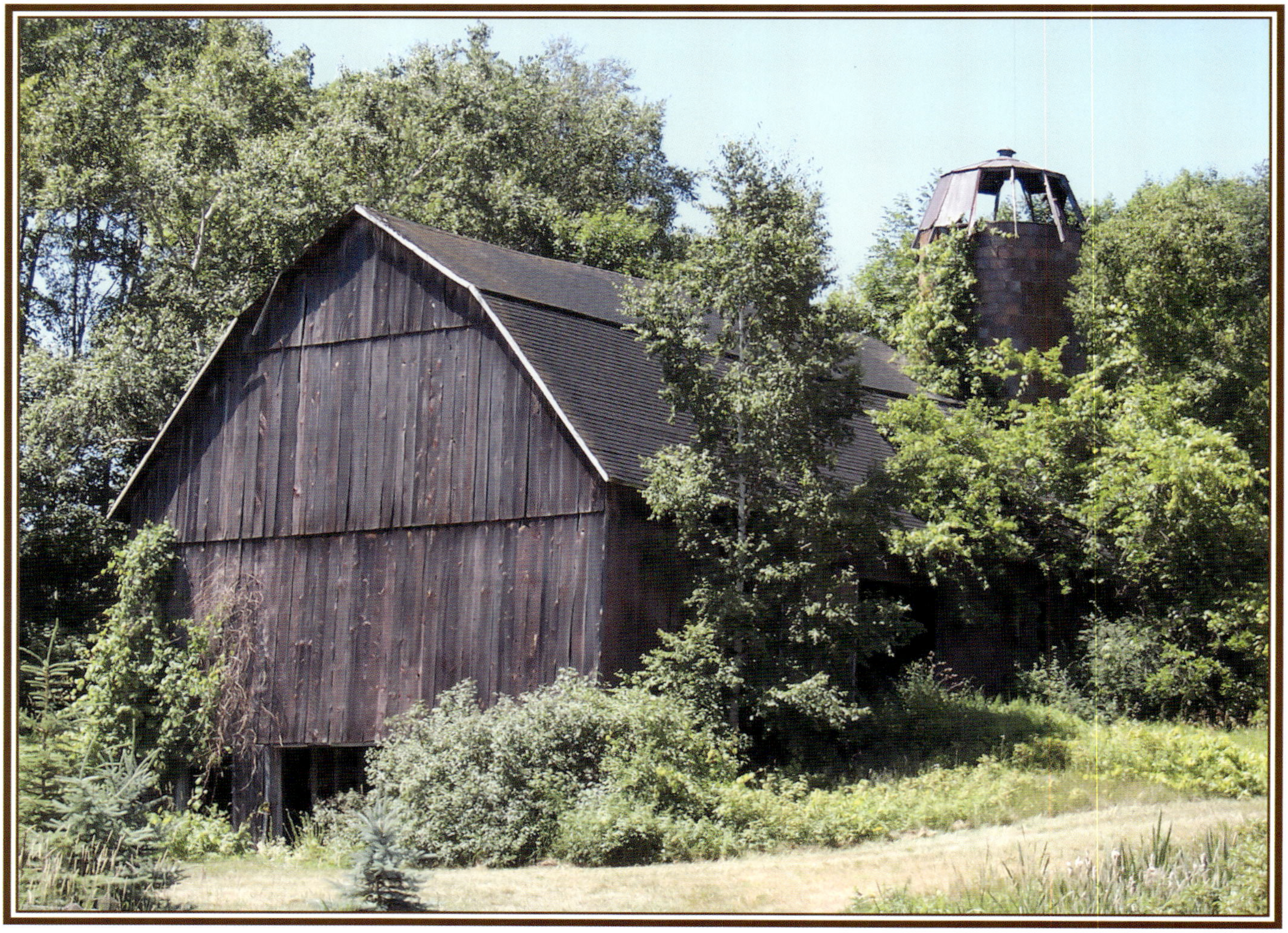

OLSEN/TIEFENBACH BARN

"In the early 1900's my grandfather 'Charlie' Olsen built a large barn on his sixty-four acre farm on Bluff Road. After Grandfather's death in 1939, my grandmother ran the farm, growing cherries until she passed away in 1943. The farm was called Two Oaks. My mother Elmira grew up on the Peninsula and inherited the farm." This is from Mary Ellen Rabine, granddaughter of Charlie Olsen. Elmira married T.N. Sproule and moved to Detroit. The farm became a gentleman's farm where the family came in the summer. Mary Ellen continues, "I have many wonderful memories of Grandpa's barn. It was large with two floors and built into a hill. The upper floor opened onto the top of the hill. There was a large double door and an opening on one wall for hay to be hoisted from the wagon and lowered into the barn.

Mary Ellen recalls, "It was a wonderful place to play. We liked climbing up into the loft where interesting things were stored, like tin cups that once contained water for baby chicks that were shipped from a mail order catalogue. We liked playing hide and seek. There was an opening in the floor where the hay was lowered for the animals—a neat place to sneak into the lower floor. Grandpa had running water in the barn, piped down from a spring at the top of a hill beyond the meadow.

"When Grandpa would unhitch the horses from the wagon that was stored in the top of the barn, he would let us ride down the hill on the horse to the lower level of the barn that stabled the animals. Along with the horses there was a cow and a calf and a separate area for chickens. I never saw pigs in the barn, but the reference to salt pork made me think they might have, at one time, raised a pig or two for butchering. During cherry season some of the cherry pickers that stayed in the barn were people from the South and even a circus person. They hung sheets or blankets for privacy."

Physical Description
Date Built:
- Early 1900's
- Moved to present location in 1948

Construction:
- Timber frame bank barn ramped with glazed tile silo
- Vertical wood siding
- Asphalt gambrel roof
- Cement and stone foundation

Around 1948 the barn was moved, the foundation of stones was bulldozed out and the land became an extension of the cherry orchard. Cal Jamieson remembers helping when the barn was taken down and moved to Uncle Bud Jamieson's farm one and a half miles north on Bluff Road. "They had an old fashioned barn raising," Cal said. "Several men would put a side together on the ground and then raise it in to place." Bud Jamieson grew about 120 acres of cherries at one time.

This 40 x 54 foot gambrel roof barn is almost hidden as it is surrounded by trees and foliage. It has an asphalt single roof and a magnificent glazed brick silo still attached to the barn. The top of wood is in need of repair. The Tiefenbachs, who bought the farm in 1986, do not use the barn but still have about 18 acres with about 400 to 500 cherry trees. Another local fruit grower cares for their orchard. This farm has had several other owners, the original owner being the Schaeffers. They grew field crops such as potatoes. Belmont Schaeffer, who perished in World War II, was the previous owner to Bud Jamieson.

Information Cal Jamieson, the Tiefenbachs, and information by Mary Ellen Rabine, <u>Barns and Blessings</u>.
Picture by Robin Grothe.

1904-05
Harry Seel, the apple man, says the condition of the trees could not be better. The buds were in excellent shape for a hard winter while the continued fall of snow has been a great protection. Mr. Seel has a farm near Old Mission of 101 acres, 95 of them given to apples alone, approximately 5,000 trees. (We live on some of Mr. Seel's orchard property now...Evelyn)

Urtel wellhouse

URTEL BARN

The beautiful 36 x 42 foot stone and wood barn was built in 1914. It is a two story barn with a stone walkout basement. The barn and the house are of the craftsman style built using a great deal of stone and exposed beams. The lumber for the barn came off the farm. A beautiful uncoarsed (no mortar showing) cobblestone wall topped with a concrete cap leads up the driveway to the house. This wall goes around part of the house and is the first thing you see as you drive in. It is a truly lovely setting, pearched on the highest point of the Peninsula near Carroll Hill, nestled privately behind a grove of trees.

The Urtels bought the six acres in 1998 from the Langleys who owned the property from 1991-98. The barn now is used to house horses in the lower level and hay is stored in the upper level. Mrs. Urtel told me that they have recently restored the barn by adding a new roof and shoring up the walls and foundation of the barn. She estimates the cost of the repairs at $20,000. Early on, the barn was used for dairy cattle and later it housed cherry pickers.

The original owner was a man named Burkhardt. Later it was owned by Dr. Gleason, and next the Minnema family farmed 130 acres for 40 years growing cherries, apples and plums.

Physical Description
Date Built:
- 1914

Construction:
- Sawn lumber placed horizontally
- Stone on lower side and brick on upper sides
- Asphalt shingle gambrel roof
- Fieldstone and concrete foundation

In the late 20's and early 30's when Mr. Frank Burkhardt owned the property, he decided to build a canning factory down the hill east of the stone barn and house. His plan was to can cherries. He also experimented with maraschino cherry processing. In 1931 he ushered in a new market and a new era for the sweet cherry. He was a most enterprising man. In 1926, the first Cherry Queen and he took a cherry pie to Washington D.C. to present it to President Calvin Coolidge. Their mission was accomplished, and they made all the newspapers with the pictures taken. You might say, "they put Traverse City, Michigan, on the map." However, many obstacles in the canning business kept him from being as successful as he would have wished. He could not access water and had to haul it up from the Bay in water wagons round the clock to keep up with the demand. Then there was the problem of getting supplies out to the site. He tried to convince others to put a railroad out on the Peninsula. This never came to "fruition." He did get a steam locomotive, using the boiler to heat the water to facilitate the canning operation. The whole operation lasted about 7-8 years. Mr. Burkhardt finally gave it back to the bank and got out of the canning business. The story is told that he came to the farm driving a big car, and when he left he was driving a team of horses. George McManus told me that he did have a farm on M-72 and continued to farm. Here was a gentleman that did not give up on farming!

There might have been a lesson here that Mr. Burkhardt overlooked or ignored. I read in the *Grand Traverse Herald* that "an enthusiastic meeting was held last week at Old Mission in regard to the proposed new railroad on the peninsula. It was met with the heartiest approval with $2000 of the required $10,000 being raised at once and half the right of way was given. W.R. Pratt, E.O. Ladd, and A.E. Porter were appointed to take charge of securing subscriptions and the remaining necessary right of way. This road would greatly reduce the estimated $25,000 annual loss on fruit and produce." This was written over 100 years ago (1904) and THAT railroad was never built either.

Information from Jill Urtel, Whitney Lyon, George McManus Jr., and John Minnema.

December 1905
The roads in the area are in very poor condition with not enough snow to make good sleighing but enough to make the farmers wish they had their corn husked and a warm place for their cattle. Our correspondent at Ogdensburg reports that Tony Gowlich, a farmer at that place, has his corn shredded at Mr. Golden's but it's no telling when he will be able to do the job on account of the weather.

GARY *and* WENDY WARREN BARN

This barn has a famous, or infamous, history, but let's start at the beginning. The Carroll family owned this 28 x 32 foot barn for many years. The 1895 plat map shows that L. Carroll owned 50 acres. Next, Stephen Carroll owned it for many years. The farm consisted of 35 acres of farmland on both sides of Island View Road. His friends and neighbors considered Stephen a most successful farmer. In 1953, Ray Carroll bought the property but later sold it to two doctors. In 1975, Gary and Wendy Warren purchased it. The Warrens now have 70 acres and grow cherries and apples. Gary Warren is a 4th generation fruit farmer. The barn, used for storage of cherry lugs and farm machinery, is a simple barn but has two levels inside with a loft. The barn has a shingled, gambrel roof with lightning rods intact and a fieldstone and stone block foundation. The barn has been painted (2005) a bright cherry red. The view from this farm is lovely at all times of the year as it sits up high on Carroll Hill.

The "infamous history" comes from the years when the farm was owned by Stephen Carroll. Stephen and his wife Mary had a daughter, Celia; Mary also had a son, Jesse Braddock, from her first marriage. In 1921, when Stephen Carroll, 52, and Mary, 47, had been married for 12 years, eighteen-year-old Jesse lived in Traverse City during the school year but stayed on the farm during that summer. He was considered to be a rather wild young man; lately he'd gotten into several minor scrapes with the law. Nevertheless, he was popular with his fellow students who reportedly liked and admired him.

On Friday, May 6, 1921 his wife found Stephen Carroll dead in the barn. He was lying on his face with his head about three inches from the stall partition. His horse, Queen, stood quietly in her stall, munching hay. Dr. F.G. Swartz was called out from Traverse City, and he arrived at the farm about the same time as Jesse Braddock, the stepson. Jesse explained that he had started out from town on foot the previous evening and had slept most of the night in a grove of pines along the shore road (now, Peninsula Drive). It seemed clear to Dr. Swartz that this was a tragic accident, and that Stephen Carroll had been kicked to death by his horse. But Queen was a gentle horse and was known to have kicked only once or twice before. John Lyon, a neighbor, wasn't so sure it was an accident. The previous night, Mrs. Carroll had told him that Jesse and his stepfather had had a great quarrel over the use of the car—Stephen had refused to let Jesse use it and had locked the garage door. Mrs. Carroll said that her husband told Jesse, "Get off the place and stay off it!" Whereupon Jesse told his stepfather that he would "get even, if it took a hundred years."

Well, testimonies were given for and against foul play by Jesse Braddock. He eventually was arrested and charged with murder. People marveled that he took his arrest with such composure and that he seemed to be a very cool customer. There was a trial, and in the end the verdict was quickly reached: "Not guilty." It seemed obvious that the prosecution had failed to prove its case. A jury acquitted Jesse, and the presumption of his innocence before the law still stands. (This version of the story was taken from Larry Wakefield's book of famous murders, *Butcher's Dozen, Thirteen Famous Murders*. The story, "The Kicking Horse Murder," is found on pages 145-151.) Another version of the story has two local men going to a fortune teller in Cadillac, Michigan, and asking what she "saw" (in her crystal ball?). She told them to follow the stream that ran across the road from the Carroll house all the way to West Bay, and they would find the murder weapon by a tree at the bottom of the hill by the Bay. It is reported that they followed up on this and did indeed find a sharp, heavy object, something like a hammer. (This version came from George McManus who heard it repeated from neighbors when he was growing up on the Peninsula.)

People are entitled to their own opinions, and a great many Peninsula residents believed that Jesse Braddock was guilty, that he had "gotten away with murder." It almost seemed fated that the young man would come to an untimely and horrific end. In 1930, as a fugitive from the law, Jesse jumped or fell to his death (some say he was pushed) from a hotel window in Wisconsin.

Mary Carroll and daughter Celia lived on the farm together after Stephen's death, though they were not always treated cordially by the neighbors after the "incident." It is said that Mrs. Carroll was a good person but was a no-nonsense lady. She once heard a prowler at her door, who would not answer her warning to identify himself. She threatened to shoot her loaded gun if he did not leave. He didn't go immediately, so she shot through the door. He left. When Mary died in 1947, Celia ran the farm with help from Harold Gray and others.

Wendy Warren, current owner of the farm, told me that the barn is still a source of wonderment for the kids and visiting friends. When asked about the death of Stephen Carroll, she said, "The version we heard was that he [Stephen] was indeed murdered but in the kitchen—beaten with a hammer—then dragged to the barn and placed behind the horse." Gary Warren farms this acreage and is a cherry grower. He also helps farm his parents' cherry farm down the road.

Information from Wendy Warren, George McManus Jr., and Larry Wakefield's book.

> *Physical Description*
> Date Built:
> - 1910
> Construction:
> - Timber frame
> - Vertical wood siding
> - Asphalt gambrel roof with lightning rods
> - Fieldstone and stone block foundation

Lightning rods or iron spire, standing two or three feet high with glass balls attached, are still on most of the barns today. They were used to conduct a bolt of lightning to the ground that would otherwise strike the barn and most likely burn it down...a real fear for every farmer. Early on they were not accepted by God-fearing farmers because they felt it was God's will for the lightning to strike, but in time, most farmers were convinced of their practicality. Many lightning rods today still appear on the barns but do not have the colorful glass balls intact. Why? It was a favorite target for farm boys with BB-guns.

Another building on the property

WELLS FAMILY BARN

This lovely barn in a picturesque setting was built in 1910. It is 24 X 40 feet. It has two levels above ground and is your typical bank barn, used for cattle on the lower level, and hay and equipment storage on the upper levels. It was owned originally by Abel and Florence Willobee. Then, in 1926, Roy and Theda Hooper bought the farm and lived on it for 50 years. They farmed…raised potatoes, fruit and had cattle. The barn was kept in good condition.

The Hoopers raised their family, worked the land, and eventually retired from farming. Roy held the responsible position in the community as the Peninsula township supervisor between 1937 and 1953. His salary was $5 a day and 5 cents a mile for services. Over the years it increased, and in 1947, he was getting $600 a year! He also was a county commissioner and chairman of the Grand Traverse Road Commission at one time. As township supervisor he had many duties. While reading the minutes of the Peninsula Township from over the years, I gleaned just some of his duties at the time. He conducted the meetings and he procured the services of people for repairs to the township hall. He had to find a suitable furnace for the township hall, find a painter for the hall, hire a mason to repair the stonewall on the hall, buy a flagpole, and see to the redoing of the inside of the hall…making a dining room and a kitchen and then redecorate it too! While he was supervisor, a garage for the fire truck was built and, later, he was in charge of overseeing the purchase of a new fire truck. (The fire company was organized in 1946.) Then, he had to find someone to shovel the snow from the fire truck garage doors in winter. He also was the health officer for a number of years for which he was paid $100 a year. Epidemics were always a threat when there were no vaccinations for small pox, measles, whooping cough, etc. Doctors were not readily available, so a health official kept track of what was "going around."

Physical Description
Date Built:
- 1910

Construction:
- Timber frame bank barn with extension on back
- Vertical wood siding
- Part metal and part asphalt shingle gambrel roof
- Foundation-part fieldstone and cement block on extension

In 1961, when he was 19 years old, Terry Wells bought the farm. As he said, "40 acres for $40,000." He eventually owned 70 acres, growing cherries, prunes, apples, and pears. Terry did not raise cattle in the lower level but used it to store his sprayer, cherry shaker, and tractors. Terry met his wife Becky, a Tennessee girl, one summer, as she always vacationed on the Peninsula. Her mother was the former Jean Holmes. They stayed with Jean's mother, Lula Holmes, on the farm. After nearly 40 years of farming, Terry and Becky retired in 2000 and now the farm is occupied by Terry's son and daughter-in-law, Michael and Betsy. Terry and Becky spend winter months in Ridgway, Colorado, but have a new home next door to their kids.

The barn is in good repair because it has been conscientiously taken care of. The original timber logs from the farm make up the frame work of the barn. The upper and side foundation is fieldstone. The back foundation was replaced with cement block around 1932, according to the date written in the cement on the floor. The metal roof is being replaced by shingles. The barn is used for storage. The property is approximately 12½ acres now. There is a flaired eave gambrel roof building opposite the barn that housed migrants when cherries were picked by hand. The migrants came to work on the farm, 50 adults and about 35 children, until 1971. Terry said they were good workers and good friends. Now Mike Wells has a ski tuning shop in the upper level and uses the lower level of this lovely ivy covered building for storage.

This peaceful setting sits high above Smokey Hollow Road and is hidden from view by the trees.

Information provided by Terry, Becky, and Michael Wells. Also from the Peninsula Township Board minutes.

1904-05
Henry Lardie, who is a merchant of Old Mission, while playing ball Saturday afternoon with a crowd of boys, fell and broke the bone in his right leg just above the ankle. Medical attention was requested from this city and a physician was sent immediately to tend the patient.

JOHN WUNSCH BARN

Picker's quarters from the past

Front of the barn

JOHN WUNSCH and LAURA WIGFIELD BARN

This property is one of the original farms on the Peninsula. William A. Marshall came in 1862 from Frenchtown, Michigan and purchased 240 acres from E.P. Ladd. "Mr. Ladd had previously cleared part of the land and built the buildings which have since been enlarged and modernized so that now they are the best on the Peninsula while the broad fields of corn and waving grain, and the work well done, all testified to the good management and thrift of the owner." In later years Mr. Marshall grew fruit. "There is a fine large apple orchard and one of cherries, while the vineyard yields more grapes than are grown anywhere else on the Peninsula." (From William Marshall's obituary, *Grand Traverse Herald*, Jan. 27, 1914.) The acreage was about 80 acres when he died.

"After he had chosen his land on the Peninsula, high on the ridge, William Marshall went back to Frenchtown for his family, which came overland, driving their cattle with them. At the end of a long day they were forced by night fall to pitch camp near their destination, but it was only in the morning that they discovered they had spent the night but a few hundred yards from their goal." (*Story of Old Mission*, Potter, p. 71)

This weathered gray, gambrel roof barn was built in 1905. In places you can still see remains in rusty red paint on the vertical siding. This 42 x 46 foot barn is probably the second barn on the sight. John Wunsch speculates the first barn would have been built around 1862 at the same time as the house. The house is still standing and has been renovated by the Wunschs. It was probably the third stick-built house on the Peninsula, according to John. Early on, the farm had 80 acres in cherries, apples, grapes. Cattle were kept in the lower level of the barn. The family grew their own hay and had a trolley system for hauling hay to the loft in the upper level. There is a large extension on the back of the barn.

Physical Description
Date Built:
- 1905

Construction:
- Timber frame bank barn, extension on back is plank frame
- Vertical wood siding
- Asphalt shingle gambrel roof
- Fieldstone foundation with cement under back extension

Picker's quarters

This timber frame barn has hand-hewn tie beams. The lengthwise beams probably came later (1900's), as they are sawn lumber. There was a silo at one time, with walls that stone, 2 feet thick. The farm has many outbuildings as there was need for migrant housing. I counted seven outbuildings, including 4 migrant cabins and a long row of concrete migrant housing on the property.

In the 1940's this farm had the biggest cherry weigh station and packing plant on the Peninsula. Farmers brought their cherries here for weighing. Then the Pratt-Gleason operation hauled the fruit to processing plants, usually in Traverse City, in their own trucks. Morgan Fruit Packing and the F & M plant were two such processing plants in town. The Peninsula station was run by Mrs. Marshall Pratt in partnership with the Gleason family. The Marshall and the Pratt families, descendents of William Marshall, owned the farm for many years. In the 1970's the farm was purchased by the Gleason family.

Receiving station

In 1986 John Wunsch and wife Laura Wigfield purchased it. John and Laura have preserved the barn by putting a new roof on it and repairing the siding. The Wunsches value the history and heritage surrounding this property and are proudly restoring it. It takes patience and cash to do this. Reading the history of the farm when William A. Marshall owned it, in 1869 he wrote a letter to his wife saying, "I just got my taxes paid. Ninety dollars and seventy-five cents. It was hard work to collect enough but I finally made out. I took some apples up to Traverse City and sold them. I got $7.50 per bushel." (*Memories Hidden, Memories Found*, p. 198-9)

Oral information from John Wunsch.

ROGER and MARTHA MYERS BARN

Charles Cooledge was in Idaho when he met Addie McManus. Her mother, two brothers, and she had left Old Mission Peninsula to homestead in Idaho, living on the property for five years. Charles and Addie eventually married, sold the homestead property, and moved back to the Peninsula. The rest of the McManus family followed soon after. Charles and Addie bought 45 acres from Lawrence Carroll on the corner of Center and Island View Road and settled on the farm somewhere between 1906 and 1908. Charles Cooledge's name appears on the 1908 plat map.

The 30 x 40 foot gable roof barn was built about 1910. It has an interesting gable dormer in the middle of a large shed extension to the main barn. This gable roof dormer sits on the long sloping roof of the extension. It has deteriorated asphalt shingles, and the barn itself is in decline. The barn is no longer used except for storage.

This farm was a fruit farm, growing apples, sweet and sour cherries, and peaches. The barn itself housed cattle, hogs, and chickens. There was a pasture on the acreage, and they grew hay for the animals. It was a self-sufficient farm. Ten acres on the northwest corner of the property was sold to the school district in the 1950's to build the new Old Mission School. Land to the north was also given to the Catholic Church to build a cemetery. Lawrence Carroll owned the property at the time. The Hoffman family also donated land on the south side of their property. More land was added to enlarge the cemetery in 1956.

Physical Description
Date Built:
- 1910

Construction:
- Timber frame barn with large front extension and dormer added
- Vertical and horizontal wood siding with shiplap siding on three sides
- Asphalt shingle gable roof, sloping roof on shed extension
- Fieldstone foundation, shed extension is cement block

Most of the information I have on this farm is from Ken Cooledge who grew up on the farm. His parents, Fred and Lillian, had come to the farm in 1933. When Grandfather Cooledge died in the 1940's, her son Fred farmed the land. Fred's son Ken told me that he has been told that the original house was close to Center Road and that it burned in 1912. It seems that the family was gathered for Sunday dinner when they heard a man, who happened to be driving by with horse and carriage shouting, "Your house is on fire!" He kept right on trotting down the road! In those days it was impossible to save a burning house once the fire got a head start. Fire departments did not exist. The next house was built set back from the road, near to the barn. Ken's mother, Lillian was known to be seen shoveling that long driveway in the winter.

In the 1950's the family got rid of the cows and concentrated on cherries. Ken remembers working in the orchards and that they hired many migrant pickers. He said he has pleasant memories of playing with the migrant children after cherry picking was done for the day. He said his brother Fred and he played hide and seek and had ball games with the migrant children. It was a good part of the summer for him, as there were few neighbor children to play with.

Ken remembers painting the barn red. But the Cooledges had a good reason for picking red this time. It seems that red paint was $5 for 5 gallons at Montgomery Wards, a bargain at the time as a barn takes a lot of paint. Over the years it has been repaired more than once. Ken said a storm in July 1956 blew off the siding on the south side of the barn, and the replacement boards were placed horizontally. Also in 1956, the stone foundation was crumbling so a cement block front was put on the barn.

Coming down Carroll Hill, it is a pleasant sight to scan the countryside and see two or three old red barns gracing the land, holding their own. One of those barns is the old Cooledge barn. After the Cooledges sold 35 acres to Al Carroll, they kept the house for 4-5 years. Later, St. Joseph Catholic Church purchased it to be used as a religious center. Now it is owned by the Myers family, who have already started to renovate the house, with the barn to follow. "We want to go back to our farming roots," said Mrs. Myers.

Information provided by Ken Cooledge and George McManus Jr.

THE BARNS OF OLD MISSION PENINSULA
1920-1994

BEERS BARN

Timber frame beams inside the barn

BEERS BARN

One Sunday, my husband and I were driving down Peninsula Drive and came upon this magnificent big red barn, nestled among a grove of trees. Crowding around it were lovely homes and expanses of carefully groomed lawns. The Beers, Mack and Lorraine, came out to talk to us, saying they had owned the property for 20 years. It was originally 40 acres but now is just two and a half acres. The Beers bought the site from Jim and Chris Cross.

The original settler was Ludwig Nelson and his wife, Jennie Swanson, who came over with a group of Swedish immigrants in the late 1800's. They settled on Old Mission Peninsula because the rolling hills and the beautiful water reminded them of their homeland. Ruth Nelson Cilke, along with her two older brothers, Philip and John Nelson, grew up on the farm. She enjoyed talking about her old homestead.

The house was built in 1888, but the existing barn came later, built by Ruth's father, Albin Nelson, in 1920. Albin was born on the farm, and he farmed the land until he died in 1968. The barn is a 40 x 32 foot timber frame with a gambrel roof. It has a raised ramp to the main level of this three-level barn. The track for moving hay still is in place. The foundation is of fieldstone, and the bright red siding is still in good shape. The large log beams in the lower level are from timber taken off the property. The lower barn door was enlarged so Ruth's older brother John could work on cars in the barn. Albin also had a cider press on the farm. He replaced his work horse with a John Deere tractor around 1954.

Physical Description
Date Built:
- 1920

Construction:
- Timber frame bank barn, double ramp
- Vertical wood siding
- Metal gambrel roof
- Fieldstone foundation

There is a foundation of an ice/milk house still in place right over a spring fed creek that runs down the hills behind the barn. Ruth remembers the building was a 20 x 20 foot miniature house and served as the ice house. Ice was cut from the Bay by her father in the winter and stored in the ice house all year long. The ice was covered with sawdust to save it from melting. When you looked for acreage to farm on the Peninsula, it was very important that it have a spring, creek, or some source of water on it, as wells were not commonly drilled in the early times.

The Nelsons grew more than cherries. They also had apples, peaches, and plums. Because the land behind the house and barn is hilly, they had plateaus of orchards, making it easier to work the farm. Ruth remembers helping with the pruning of peaches. She said she used a rubber hose to beat off the extra peaches, thus allowing the remaining peaches to grow bigger. Ruth remembers the "colorful" migrant workers who came to help with the fruit harvest. One was Huey, a hobo, who showed up every year at harvest time. He preferred to live way back in a shack, wanting to be alone. Another worker, Tom, had a peg leg.

Ruth has many pleasant memories of living on the farm. She said that one Halloween she invited her Campus Life group to a haunted house party in the barn. They started in the orchard out back and came into the barn in the dark, where all sorts of spooky antics took place. Her best friend, Holly Haven Elliot, lived at Neah-ta-wanta, and the two of them loved to play in the barn, swinging from ropes that were hanging from the rafters. As they would swing back and forth, they made up commercials for whatever they could think of. Ruth also remembers being in the first class of the new Old Mission Peninsula School when it opened in 1955.

In the winter, Ruth loved to go down to the Bay and walk on the ice. It was very quiet, she said, except for sounds coming from the ice. She would use her imagination to think up all sort of things that might be causing the sounds. She said, "I cannot think of a better place to grow up. I had parents who loved me, lots of fruit to eat, and a bedroom window that overlooked the beautiful Bay. I always felt blessed."

One final story told by Ruth: Her grandfather had a special box of carpentry tools he had brought with him from Sweden. It was cherished by him, and later, his son, Albin. It seems a local lawyer and avid antique buff saw this box of tools in the barn and wanted to buy them. Albin was not interested in selling and, fearful that they might be taken somehow, he hid them somewhere in the barn or in one of the other buildings on the property. The problem was, he never could find them when he wanted to use them. The whole family searched for that box of tools repeatedly, and they never were found.

Information provided by Mack and Lorraine Beers and Ruth Nelson Cilke.

DEBRA BEE BARN

DEBRA BEE and RUSSELL PATTON BARN

Manasses Swaney was the seventh child of John and Rosannah Swaney, among the earliest pioneers who came to Old Mission. Manasses received 80 acres across Swaney Road from his parents' homestead when he came of age. Manasses married Mary Ellen McGarry and together they had three children. Two sons, Charlie and Rob, each received roughly 20 acres when they came of age. This was land adjoining their parents on the east side of Center Road. Fannie, their daughter, never married, living in the house the rest of her life. However she did inherit the 37 acres her parents had on East Bay on Smokey Hollow Road. Her mother, Mary Ellen, died in 1941. After Fannie Swaney died in 1947, the farm was purchased by Ted Ayers, a Swaney relative, and now Debra Bee and husband Russell Patton own the farm, which is 6.75 acres.

Not much is known about Manasses Swaney, but one good story tells how he left this world with a vivid remembrance by all involved. Manasses died when he was 56 years old in 1900. In those days it was the custom for several members of the family to "sit up" with the dead, that is, stay awake all night in the same room with the body. Being an unseasonably warm night, a window above the table where Manasses was laid was open a crack. Along into the wee hours of the night, the men keeping watch started to nod off and tilt back in their chairs. About that time it seems one of the geese Mary Ellen kept decided to stick its neck into the window and let out a horrendous bellow. It is said that all three men got stuck in the doorway trying to get out of the room! (From Swaney Family History, compiled by George Beckett)

Physical Description
Date Built:
- 1929

Construction:
- Timber frame on grade barn
- Shiplap horizontal wood siding
- Asphalt shingle gabel roof
- Foundation is cement footings

The barn is of timber frame construction and was built in 1929 by the Swaney family. It has shiplap horizontal siding. The foundation was hewn wood beams set on large rocks, but the current owners have put cement footings around the barn to shore it up. The roof is asphalt shingles over wood shakes. There is a standing rib steel roof on the side shed. This part of the pretty little barn is now a stable for the horses. Hay was hoisted from the ground outside into the loft. The barn is used for storage now, but it also houses many nesting swallows, raccoons, and cats. (At one time it housed migrant workers too.) The farm had cattle in the past, but they were kept in a barn beside the existing barn. It is a beautifully preserved barn, trimmed in white with a new, easy-care metal door for getting tractors in and out. Debra Bee, a relative to the Ayers, has lived on the farm for 20 years.

Information provided Russell Patton and Debra Bee.

June 22, 1905
Early Thursday morning last, the home of Amos Sweet of Old Mission caught fire and burned to the ground in spite of the hard work done by a bucket brigade which took water from the Bay. All of the furniture and practically everything else was saved. The cause of the fire was a defective chimney.

Old barn before moving and renovation - 2001

Summer 2001
Large barn up by pole barn. Mover getting ready to relocate it in the valley.

Summer 2002
Large barn in new location, being primed and painted.

BRYS FARM

This simple, single-level gambrel roof barn stands back in the field dotted with 23 acres of grapevines. It adds a splash of red to the otherwise vivid green landscape in the summer on our lovely Peninsula. This 30 x 44 foot barn was moved to its current location because it was too close to the farmhouse, a lovely old house that has been completely restored by Walter and Eileen Brys (rhymes with wise). The Bryses bought the 80-acre former cherry farm in 2000, with the dream of starting the fifth winery on the Peninsula. Now it is called Brys Estate Vineyard & Winery, and was opened in 2005. Another small barn was converted into a guest house behind the main house, and in years past was used for migrant quarters during cherry season. Beautiful gardens surround the houses in the summer. The setting is totally breathtaking as you can see glimpses of East Bay from their high ground.

Of the first 40-acre parcel, 23 acres are now planted in grapes. The second 40-acre parcel is planted in rye, enriching the soils for future vine planting. The vineyard is harvested completely by hand instead of by machinery. It is considered to be a relatively small vineyard at this stage. But being small has its advantages. All of their wine is produced from their own grapes.

In the past, the farm had orchards of cherries, apples, and peaches. The first owners were the Brown family. One of the Brown daughters married a Giles, and the farm became known as the Giles Farm. The Giles lost their barn to fire and rebuilt it in the 1930's. Arnold Elzer bought it from the Giles in the 1970's. Now it is just a pleasant sight, sitting way back in the fields.

Most information provided by Eileen Brys.

Physical Description
Date Built:
- 1930

Construction:
- Timber frame on grade barn
- Vertical wood siding
- Metal gambrel roof
- Cement block foundation

FAR OUT FARMS

It would be difficult to forget the Far Out barn once you have seen it, as the name is spelled out on the asphalt roof. The barn with a gambrel roof is used mainly for storage. When I peeked in one day, the sweet smell of the hay stacked on a wagon was wonderful. The barn was built by Roy Hooper, Jr. around 1943. It was used as a dairy barn first by Roy's father. The roof containing the words "Far Out Farms" was added in the 1970's by Frank Maveedy. The farm has had a number of owners, but let's start way back at the beginning.

Enoch Wait came to Old Mission in 1855 as one of the first settlers to the area. Enoch was a carpenter, building his workshop on the Bay by the general store and the church in Old Mission. This was before all the buildings were moved up from the Bay. He also had one of the first houses along Ridgewood Road. The 1880 plat map shows a house at the front edge of his rectangular 120 acres. This house still stands today, and is owned by the Jamie Marsh family. The township records show the house was built in 1870, and when the Marsh family was renovating the old house, they found the name E. Wait written in beautiful script handwriting in two places in the walls. The Marshes' land consists of $9^{2}/_{3}$ acres. They purchased it from the Cavender family in 1999.

Physical Description
Date Built:
- 1943

Construction:
- Timber frame on grade barn
- Vertical wood siding (cement block wall on west side for milking parlor)
- Asphalt shingle gambrel roof
- Cement block foundation

According to *Sprague's History* (1903), "Mr. Wait purchased a claim from the government. The land was wild and unimproved, not a furrow had been turned upon it, but with characteristic energy, he began its development and in the course of years he transformed the wild tract into highly cultivated fields. He has made a close study of the needs of the kinds of fruit which he produces and his opinions in such matters are largely received as authority by his neighbors. He follows progressive methods in all his works" (p. 753). Enoch Waite was highly respected in the community and was almost 99 years old when he died.

The first trees (apple) were planted in this area by the Chippewa Indians. They knew the growing conditions, soil, and climate that were favorable long before white men came (from *Memories Hidden, Memories Found* p. 105).

Eugene Wait, Enoch's son, was born in 1859. He carried the mail from Traverse City to Old Mission for seven years. When he married Jennie Phanstat, they built a house next door to his father's and cultivated 80 of their 100 acres in fruit trees. (From *The Story of Old Mission*, by Elizabeth Potter, p. 74). This house is dated as being built in 1885. It is now owned by James and Karen Olsen and Evelyn Lundeen.

Martie Hyslop remembers Mr. Gene (Eugene) Wait had horses in the barn, and on snowy days he would wait to get his mail in "town", (Old Mission) until it was time for the kids to get out of school. Then we would pile them in his sleigh and drop them off at their homes on his way home. He spent most of his long life on the Old Mission Peninsula.

According to Dr. Potter, "One of the other owners on this stretch of property was Captain Ace Johnson, who came to the Peninsula between 1881 and 1895. He first settled up on Ridge Road, buying land from Sol Franklin. He was a Norwegian seaman and captain of an ore barge on the Great Lakes. Captain Johnson later purchased the Enoch Wait house. After his death, children in the area believed they saw his ghost when passing the house late on dark nights. This impression was probably due to the clump of birch trees in front, which would certainly create a ghostly impression on a windy night with their waving branches" (From Potter book, p. 78).

Captain Johnson's daughter married a Mr. Kirkkoff and they lived in the house in the 1930's until she was widowed. The property was sold to Jim Bryant in the 1940's to 1950's, who owned about 11 acres. Fern Bryant turned a small barn on the property into a resturant. She was ahead of her time as she featured a single entree…she started out serving only waffles. Later, she served only steak. You picked out the steak you wanted from a meat case for Fern to grill. It came with other side dishes and always ended with a cherry tart. The resturant had a Viking theme and the waiters were all male. She also had a gift shop in the old barn, where you could while away the time waiting to be seated in the little barn resturant. The barn burned in the 1950's, but bits and pieces were salvaged and what was left was incorporated into a pretty little barn that stands behind the Marsh house today. The Bryants sold the property to Earl Wysong sometime in the 1960's.

The Bryants moved to the other side of the Peninsula and bought a large home on Bowers Harbor that had once belonged to the Stickney family. They started a classy resturant called Bowers Harbor Inn. In the late 1960's Schelde Enterprises bought the resturant from other owners after the Bryants, and completely redecorated the interior, turning a six-car garage behind the big house into the Bowery Resturant. The rustic barn theme of the interior of this resturant includes lumber taken from the Tom Hoffman barn on Center Road when it was no longer needed (listed under Phantom Barns). Incidently, the ghost of Mrs. Stickney, who reportedly took her life in the Bower's Harbor house long before it became a resturant, is said to be seen occasionally in a mirror in the stairway. This is just a cautionary note. The property has now been purchased by Jon Carlson and Greg Lobdell and is due to undergo another renovation.

Other owners on part of this stretch of land on Ridgewood Road include the Andrus family from 1963-1972, who bought the barn and acreage from the Hoopers. Maveedys owned the property next in the 1970's, and the Hagelsteins in the 1980's. Enoch Wait's 120 acres have been divided and reorganized a number of times. There have been other owners also, but the Far Out Barn and acreage is owned presently by Logan Schlipt.

Information provided by Jackie Burns, Jon Andrus, the Jamie Marshes.

JAMIE MARSH BARN

CUTLER BARN

The Cutler family moved north from Grand Rapids in 1994 with their large family and built a lovely, sprawling country home (14,000 square feet) on the Bay. It is inspired by turn-of-the-century Victorian homes and looks like it has been rooted in place for a generation. The homestead sits on a 600-foot frontage of the Bay.

The Cutlers have nine children, now all adults. Dick and Carol promised that each child could have an animal on their new farm when they moved north. Some of the children were married and had left the "nest," but for those still at home, the Cutlers were going to carry through on their promise. They had come from city living and were anxious for space—to be country kids. One daughter, Katie, especially wanted a horse. So...a beautiful, big red barn was built to house horses and sundry other animals. This barn is one of the most recently built on the Peninsula. However, before the dream could be realized, Carol was diagnosed with a serious form of cancer and could no longer handle the work load that larger animals required. The Cutlers also ran the local supermarket at Mapleton. So it was decided that the four kids still living at home could each have a dog. At one time, there were four dogs running to and fro on the "farm".

Physical Description
Date Built:
- 1994

Construction:
- Plank frame barn
- Vertical wood siding
- Steel gable roof with cupola, lightning rod on the front
- Concrete foundation

The barn sits on the other side of the road from the home, which is a B & B, now that the children are pretty much out of the house. (By the way, the house has ten bedrooms and ten bathrooms.) There are about 23 acres of property behind the farm. In the lower level of the barn are four beautiful horse stalls all constructed and finished in oak. On the main floor is a NBA regulation basketball court for the kids' recreation and fun. The inside walls of the barn are cedar. In the loft, the floor is meticulously stenciled with an 1800's print in subtle colors done by Richard Cutler. There is a TV and lounging furniture in the loft, making it one very special place to while away some time. It is a beautiful setting, and I mused that the setting, the house and barn have more the look of an age gone by than do some of the authentically old homes and barns on the Peninsula.

Information provided by Carol Cutler.

FEIGER BARN

While the records at the township list this barn as being built in 1940, there is evidence that there was a previous barn—a date of 1910 is written in a cement and stone corner of the barn down in the lower level. This is a three-level barn and has been cared for very nicely. A Mr. Wystrand built the barn and the house as a young man. Walt Feiger told me that Mr. Wystrand came back to see the farm when he was in his 70's.

At one time in the past, it was the Erickson farm, I have been told. The Feigers bought the property in 1965. Susan Feiger and her mother started an antique business on it in 1966. It all started when they acquired a great many old things from Susan's grandfather's house. They had planned to just have a Barn Sale, but after the sale was such a success, the decision was made to turn it into an antique shop. A basketball court for their sons had been planned for the old barn, but after the sale, the basketball court idea went by the wayside and it became "This Old Barn Antiques." Later it was called "Walt's Antiques."

Much was done to improve the barn. It was reroofed in 1993 and the double loft was stabilized with cables. Skylights were added to bring more light in, and a new ventilation system was put in. On the main level the floor boards of beautiful 10-12 inch yellow pine remain. Windows and stairways were added too. Outside, the pine plank siding has been left natural wood color and the fancy roof vents have weathervanes in the shape of arrows on them. Walt told me the weathervane and arrows are 94 years old. The basement is warm and snug because the walls are of thick stone.

Information provided by Walt and Susan Feiger.

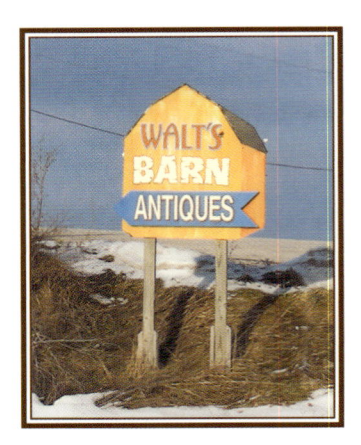

Physical Description
Date Built:
- 1940, evidence of an earlier barn

Construction:
- Timber frame bank barn
- Board and batten siding
- Asphalt gambrel roof with 2 ventilators on top
- Stone and cement foundation

GERMAINE BARN

This is a lovely setting with a bright red barn, neat white house, and other restored buildings. Charles Germaine bought the two-acre property in 1980. When I asked him why he put so much effort into restoring the barn, Mr. Germaine said, "The barn was nice looking, and I wanted to preserve it." Music to my ears! Built in 1920, it is used for storage now. It has a gambrel roof with a shed extension on the back. The main barn is 38 x 30 feet. It is a three-level frame barn with braced rafters. The stone work on the ramp going into the main level has been replaced.

Among the first owners was Floyd Gray. Next his daughter and husband, the Lardie family, owned the farm for a long while. When Dr. Fulmer bought the property, Lewis Gray helped farm it. The Fulmer family has retained the cherry orchard up above the steep hill behind the barn, and it is presently being farmed by Karl Fulmer.

Mr. Germaine did not know much about the barn's past but did know that migrant workers were housed in it. The windows were most likely added for increased light and ventilation. He did tell me that he knew the difficulty the previous fruit growers had in getting water to the top of the hill for spraying the cherry trees. He said the sprayer would be backed down to the Bay and filled with water. Next it was driven to the bottom of the hill, and the water was emptied into a tank. The water was pumped to the top of the hill into another tank. Then, the sprayer was taken up the hill with a special rig and filled with the water pumped up previously. Now they were ready to spray the orchard. It seems like an impossible task, but many of the area fruit growers have similar stories.

Information provided by Charles Germaine and Barb Gray Springer.

Physical Description

Date Built:
- 1920

Resided:
- c.a. 1980 wood shiplap siding underneath

Construction:
- Frame bank barn with large addition on back
- Horizontal metal siding
- Asphalt gambrel roof
- Concrete and stone foundation

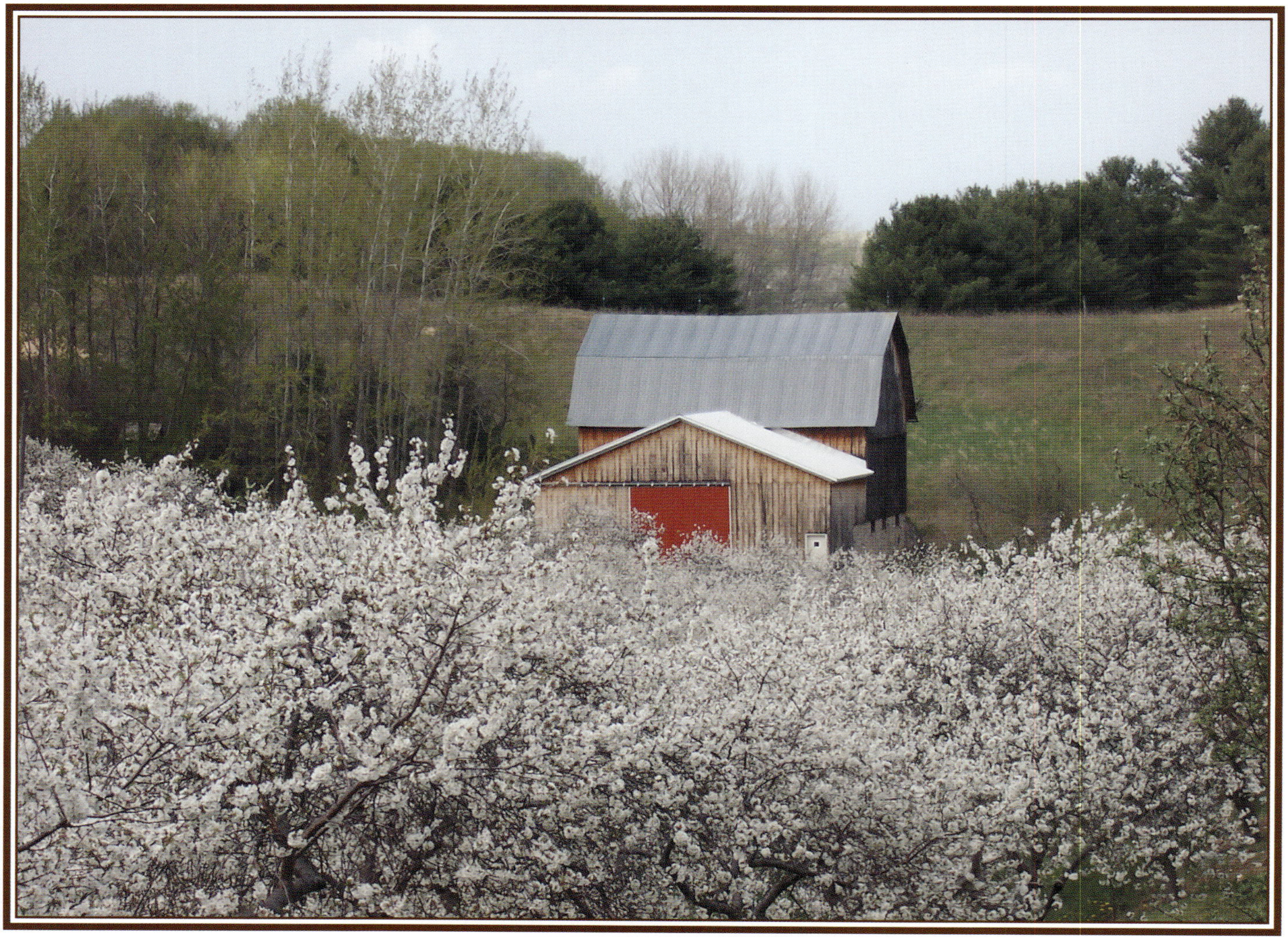

DAN FOUCH BARN

This 1928 barn was called Griffin's Hollow when Jim Griffin owned it. It sits down the hill from Center Road on Smokey Hollow Road. It had been in the Helfferich family before the Griffins bought it. Lena Griffin married Mr. Halstead, and together they farmed 75 acres of cherries. Lena farmed it herself after her husband passed away until Dan Fouch came to work for her in his college years. When the property came up for sale in 1972, Dan was a senior at the University of Michigan. He managed to arrange financing and he bought the farm from Lena.

Dan is a fourth generation cherry grower. The Fouch family had a 50-acre cherry farm on Bluff Road. It was only natural that Dan would take up the cherry growing business. Dan's father and mother, Howard and Ann Fouch, live on the original farmstead on Bluff Road. Going back one more generation, Clifford and Grace Fouch lived on and worked the Bluff Road Farm. Grace passed away in 2005 at age 101. When Dan's father retired from the cherry business, Dan bought the 30 acres from the Bluff Road Farm and 20 acres on Smokey Hollow near the Ogdensburg Church. He now farms 125 acres. In addition, Dan has been a teacher for the Traverse City Schools for over 30 years and is Director of the gifted and talented math program.

This gambrel roof barn is 26 x 42 feet and has a gable roof extension added in 1970 which is 30 x 40 feet. This was needed to house the bigger equipment used in cherry growing and harvesting today. In the past, migrant workers used the barn for living quarters during the cherry season. The barn remains unpainted although it sports a bright red door. In the past, cattle and Tennessee walker horses were raised on the farm. The cattle were housed in the lower level. The cistern in the lower level is 10 feet long and 8 feet wide on one side of the barn. This provided water for the house from the barn roof. Water from it is also used for the sprayers. All pumps, etc. were in the lower level as that was below ground and ensured they would not freeze.

Information provided by Dan Fouch and Howard Fouch.

Physical Description
Date Built:
- 1928

Construction:
- Timber frame barn with large ramped addition on front
- Vertical wood siding
- Metal gambrel roof on main barn. Metal gable roof on addition with lightning rods
- Concrete and stone foundation

Every fruit farm has a water tank used for spraying the trees

FOWLER FARM

During World War II, a Peninsula newsletter full of folksy news was sent to all the servicemen on the Old Mission Peninsula. The following was written by Carol Murray (Mrs. David Murray) in the Sept./Oct. newsletter in 1944. "The first thing that strikes our eyes as we leave the fork in the road and dodge Mrs. Trimmer's guinea hens is Frank Mathison's new barn. Frank thought he was going to join you all but the army said no, so he is going to raise some good beef for you. His dad-in-law, Ben Konig, is helping him build just the latest thing in barns, and it is up to about the third story now, and it sure looks like a 'dilly.'"

Lavonne Mathison's father, Ben Konig, was a local builder in Traverse City and had constructed many large buildings, including schools. Lavonne said she thought he built Glenn Loomis School as well as the building that houses Horizon Books in downtown Traverse City.

It is truly not your typical Old Mission Peninsula barn. It is a large monitor roof barn. Essentially, the monitor is a cupola stretched the entire length of the barn roof to improve ventilation. Modified from a true monitor style, louvers were put in this barn where there were windows in others. This barn has 4,259 square feet and, if not so old, it certainly is impressive.

Many of the original materials used in the barn were "leftovers" from other building projects. The entire framework of the barn is made of steel. The exterior is a white masonry shingle. The barn also has an unusual gable end entry. The main floor of the barn has a sub floor and on top of that is tongue and groove maple flooring. The current owners, Chris and Colleen Fowler, have had it sanded down and have a lovely bowling alley floor in the barn…complete with pins and balls set up and ready to go.

Physical Description
Date Built:
- 1944

Construction:
- Steel frame barn with framed walls, with two cement block additions on back
- Horizontal masonry shingle siding
- Asphalt shingle roof with ventilating louvers running the length of the peak and a hay hood at the front
- Poured cement foundation

It was primarily built as a dairy barn, housing, at one time, about 30 milking cows. It has a hay hood with gable end entry doors. Loose hay was lifted with hay forks connected to an overhead hay track at the top of the barn. It was dumped into the mow, when a rope was tripped that was attached to the hay track. The barn has a trap door on the main floor through which the hay could be dropped to the lower level. This level has a cement floor where stanchions are still in place. Mrs. Mathison remembers backing up a wagon full of hay to the mow door many times in her years on the farm.

It was once a working farm of forty acres, growing cherries, apples, peaches, and apricots. Lavonne ran a small store in one of the cement block buildings attached to the big barn, selling eggs, chickens, cream, raw milk, and produce. When it was no longer profitable to raise dairy cattle, Frank turned the back 35 acres into a subdivision. It has had several other owners before the Fowlers bought it, but the barn has virtually remained unchanged.

This property has been known as the Mathison Farm even though it has had many owners. The Housdings lived there before the Mathisons. The house is well over 100 years old and has been completely remodeled inside. When the Mathisons bought the 40 acres in the 1940's, the original barn was in such disrepair that it had to be torn down.

It is now a fun farm of four acres. The grounds of the Fowler Farm are beautifully landscaped. There is an inground swimming pool and a lovely guest house/office in a place that once housed chickens. The barn has been the place to build floats for many parades and the Fowlers have held many theme parties in the barn. They plant a good sized vegetable garden each summer, and have a flock of chickens, two pygmy goats, and two cats. Chris and Colleen, who just happen to be our daughter and son-in-law, even show outdoor movies on the wide expanse of their big white barn in the summer.

Information provided by Colleen Fowler and Lavonne Mathison.

Office/guest house, formerly a chicken coop

The barn before restoration

LUCIE HELLER BARN

This farm was originally owned by the Helfferich family, John Helfferich coming to the Old Mission Peninsula in 1855 with his wife Charlotte. He cleared the land and lived on the property for 34 years before his death in 1889. The Helfferichs had four sons and two daughters who lived near their parents on the northern part of the Peninsula. His obituary in the *Grand Traverse Evening Herald*, May 23, 1889 praised him, stating, "Mr. Helfferich was an honest, hard-working man. In his dealing with his neighbors he was strictly honest to a penny. We have been acquainted with him for 25 years and in all that time we have never seen him but once away from his farm."

Son John Jr. was a carpenter/builder, and he is noted in the *Grand Traverse Herald Weekly* on February 5, 1891, as building a new house for John Lardie. On March 16, 1893, Hoffman and Helfferich have the contract to building Oliver Lardie's house.

Another of John Sr.'s sons, George, seemed to have a great deal of bad luck. This news appeared in the same weekly newspaper on May 16, 1889. "George Helfferich was severely hurt one day last week by the falling of a tree." A year and a half later the newspaper reported, on February 5, 1891, George Helfferich had his foot cut quite badly in the lumber woods," and later in the month on February 19th the newspaper reported that George had his toe cut off.

Physical Description
Date Built:
- Before 1940

Construction:
- Plank frame barn on grade
- Exterior rough plywood siding
- Asphalt shingle gambrel roof
- Cement foundation

Other members of the Helfferich family have lived on the homestead including a Helfferich daughter, Mable, who moved up to the farm during the Depression. She and her husband, Pat Shavey, farmed the land. Henry, another son of John Sr., had a son, Archie, who was a township road commissioner and had a gravel hauling business. The 1895 plat map shows four parcels of land owned by various Helfferich family members, each parcel being well over 100 acres. A good deal of the land, from Center Road to East Bay, that was once part of the original Helfferich family farms is now owned by Dan Fouch. There also have been other owners of different parcels of the Helfferich land through the years.

The pioneering family grew potatoes and took them to Elk Rapids by boat, docking their boat off Boursaw Road. On the way home they would look for a very large maple tree on their property to spot where they needed to come ashore. The family also grew cherries, apples, and had the largest apricot tree in Michigan, according to Lucie Heller, the present owner.

The original barn burned, and the present barn, formerly owned by Lucie Heller, was moved from Bluff Road in the 1940's. It was built at an earlier date, torn down, and rebuilt on its current site. It is a 26 x 34 gambrel roof barn renovated by the Hellers when they moved to the farm in 1988. Lucie told me the house and barn were practically falling down when it was purchased. The barn was stabilized and reinforced with cables and the roof was replaced. New windows and a door were installed. New siding was put on over the old and it was painted a soft mulberry green as Lucie wanted a barn that "did not look like all the others." Lucie told me, "The barn was completely redone and cost $35,000 to do it!" Inside the barn there is a loft. Also, the gutters from the days of keeping cattle are still on one side of the barn. Writing carved into a partition in the barn is from the days when the barn housed cherry pickers. Old, worn wood on a part of the floor shows how much use the barn had from tractors, perhaps horses, and other equipment. Lucie uses the barn now to house her big green John Deere tractor. It is a most attractive barn and the outside belies the fact that it is a 60 year old structure.

The 8 acre piece of property goes from Center Road to Smokey Hollow and is a peaceful spot. Lucie keeps a lovely yard and has a large vegetable garden in summer. She likes the wild life, fox and the other animals that come by. She treasures the nature around her and said, "I see the natural beauty of the world sliding away from us." In 2006 Lucie sold her beloved property and moved to town.

Information provided by Lucie Heller, Tug Boursaw, and Fred Dohm.

Migrants' writings inside the barn

A raised barn with beautiful cut granite blocks with "pointed" mortar

ADOLPH and THELMA KROUPA BARN

Ade Kroupa has a truly lovely barn that is extemely well cared for, but then, Ade and his son Bill take very good care of all their property. Their pride of ownership shows. The barn is 30 x 40 feet and was built in 1937 by Bob Seaberg. The barn has a metal roof, freshly painted, and the Kroupas, concerned about the peeling paint on their barn, are thinking of having it sided in metal. You do not see the condition of the paint from the road as you drive by, but you do admire the beautiful stone foundation which has distinctive "pointed" mortar seams that bind the beautiful cut-granite foundation blocks. This makes it truly unique. Ade told me that each rock was hauled up from the fields by horses. The stone mason was Garfield Gary. This is a raised barn meaning the stone foundation is unusually high as the barn has no lower level. The Kroupas said the stone walls helped keep the cattle warmer in winter and cooler in summer. The barn is now used for storage, as there are several larger metal buildings on the property that are used for machinery and storage. Ade is a fourth generation farmer, and he has always been a farmer. Ade and son Bill farm 45 acres in cherries and have 60 acres total. Their land goes from Center Road to East Bay.

Ade Kroupa's grandfather was John Kroupa who was born in Bohemia, and was the father and grandfather of many of the Peninsula Kroupas. John had 18 children and as his sons came of age, they were given adjoining farms. John Kroupa's son, Bert, was a progressive fruit farmer and his son Ade followed in his footsteps, as did Ade's son, Bill. They have always been cherry growers. Bert Kroupa was always called Frisky Bert because of his high energy. Also, there just happened to be another Bert Kroupa over on Neah-ta-wanta Road (son of John's brother, Charles) and he was always called Harbor Bert as he lived near Bowers Harbor. Bert, Ade's father, passed away when he was 87 years old and was active 'til the end.

Physical Description
Date Built:
- 1937

Construction:
- Timber frame type, beams are laminated
- Vertical wood siding
- Metal gambrel roof
- High foundation split fieldstone with combination pointed seams and machine granite

Ade tells the story that his father and his brother Julius were very close. Bert and Julius walked to school which was at the end of Kroupa Road, but sometimes they would keep on walking right past the school and go out fishing on the Bay instead, playing hooky for the day. Bert never got over his brother's death at the age of 18. Julius had to have his leg amputated due to gangrene setting in and he died from complications.

The barn originally housed Guernsey cows. Thelma, Ade's wife, remembers Ade's mother bottling the milk and selling it to the cherry pickers. Thelma also said that on Saturdays, Bert and Anna would go to town to sell the cream to the dairy and to shop. On Sundays, they went to church, and would go visiting in the afternoon.

Thelma said. "We had about 60 migrant workers in cherry season until we bought the first cherry shaker in 1972." Ade remembers that 2002 was a bad cherry season, much like 1924 when there were none! They always had a very large garden to sustain them. Thelma laughingly remembers, "It was almost too big because we always had too much produce even when we canned and canned. We just gave it away."

The name Kroupa came from the Bohemian language meaning "barley" or "grain." Some Kroupas prefer to be called Czech, not Bohemian, as it could be considered that the word suggested gypsy or wanderer. There is a small, almost hidden, cemetery just off Neah-ta-wanta Road called the Bohemian Cemetery with more than a few gravesites, many of them with the Kroupa name on the head stones.

Information provided by Ade and Thelma Kroupa, and The Descendants of Simon Kroupa by Robert Ellis Schrader.

Electricity for the Peninsula came as far as Eimen Road in 1922, just north of Peninsula Fruit Exchange in 1932, to Old Mission Road in 1946, and to Neah-ta-wanta area in 1951.

Cupalo, with rooster weather-vane,
used for ventilation

Hay hood

LARIMER BARN

This lovely barn, amazingly, was built in Blair Township in 1937. In 1994 when it got in the way of progress, in the form of a newly created subdivision, it had to go. The land was then owned by Prevo's Supermarkets. The Larimers, Russ and Deb, saw it advertised in the paper—"Barn to be torn down"—and decided to give it a second life on Old Mission Peninsula. It was quite an undertaking. The mover, James Weaver of Norris Construction Company, took down the upper level board by board and numbered them all. Then the lower level was cut apart in twelve foot sections and numbered. In this way it was moved. The metal roof, siding, and about 90% of the components are original. It has two large ventilators, a hay hood, and track under the hood. It also has a flaired eave metal gambrel roof. Russ Larimer told me that the large barn, which is 50 x 50 feet, has 13 doors and 30-some windows. Also, he said that the movers told him it was an absolutely square building when they came to take it down.

When it was built up again on the Larimer's property, the loft was cut out in a U shape to accommodate the restoration of a very large sail boat. Russ told me that the barn was free for the removing, but there was a contingency to getting the barn. The original cement floor that the barn stood on had to be removed by the recipients of the barn, and that cost $994. The Larimers estimate that $25,000 has been spent to reconstruct and restore the barn on its current site. It is a plank frame barn built with commercial milled dimensional lumber. They have put in new windows, painted it a gleaming white with dark green trim, and it now sits tucked back behind the trees and foliage on Blue Water Road. To complete the pastoral setting, a friendly horse named Harley, makes its home in the back of the barn and can be seen in the fenced in pasture out back.

Physical Description
Date Built:
- 1937 in Blair Township

Construction:
- Plank frame on grade barn
- Horizontal wood siding
- Metal gambrel roof with flared eave, hay hood and ventilators
- Concrete foundation

The barn stored a good deal of electrical wiring and equipment when the family purchased it, as the barn's previous owner was Ben Curtis, who owned an appliance store in Traverse City. Prior owners include the Cronnenweth Dairy and George McManus, Jr., who owned the land the barn was on. The Larimer family has lived on this site since 1990 and have 15 acres. It was originally part of the Charles Lardie farm, who in the past had raised mink on this acreage. I believe you could call the Larimers a family on the move! It seems the Larimers house was also moved to the location from south of Blue Water Road on Center Road. It was one of a pair of houses built alike, sitting side by side on the Charles Lyon Farm. When one was no longer needed, the Larimers had it moved to their farm site.

Information provided by Russ and Deb Larimer.

July 6, 1905
Old Mission celebrated the Fourth in a grand manner. The festivities included a basket picnic, horse races, foot races, sack races, boat races, greased pole and tight rope walking. A big dance in the K.G.T.M. hall concluded the day's program.

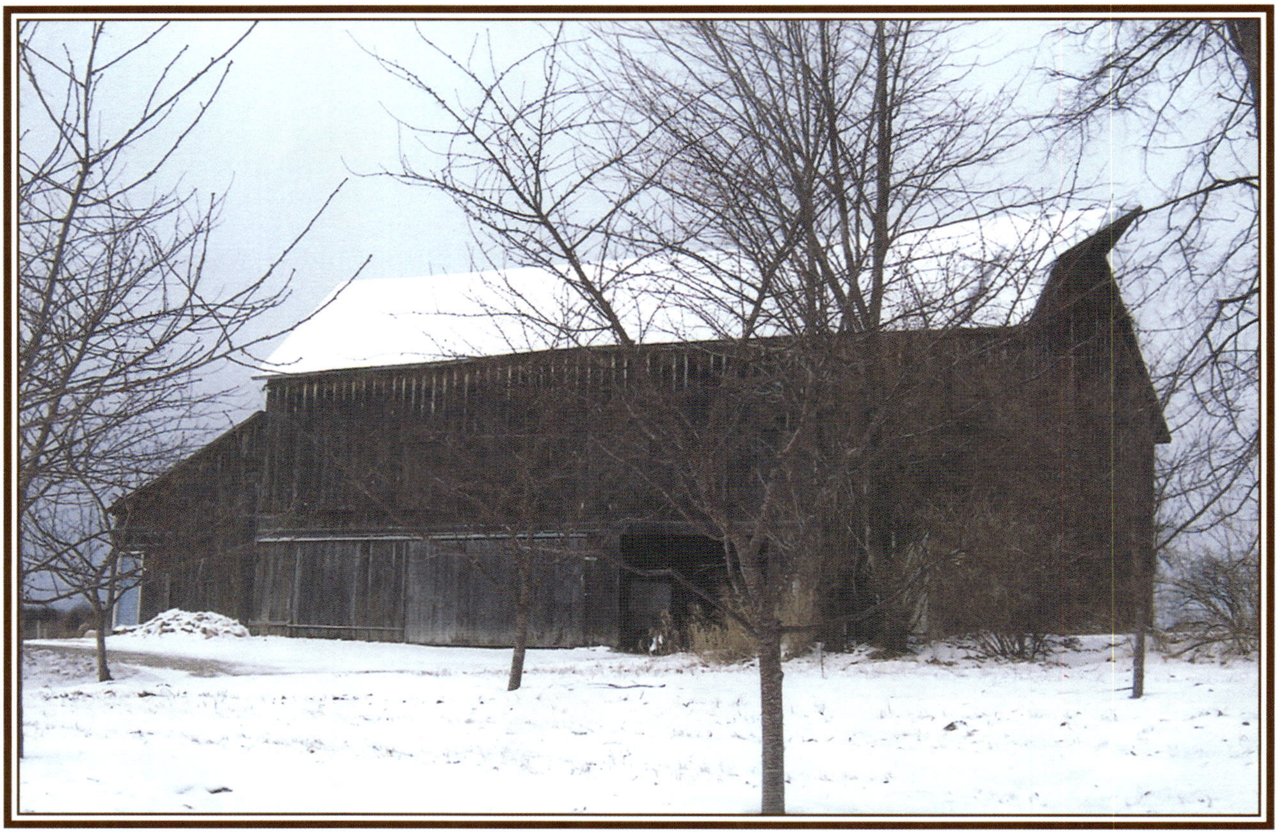

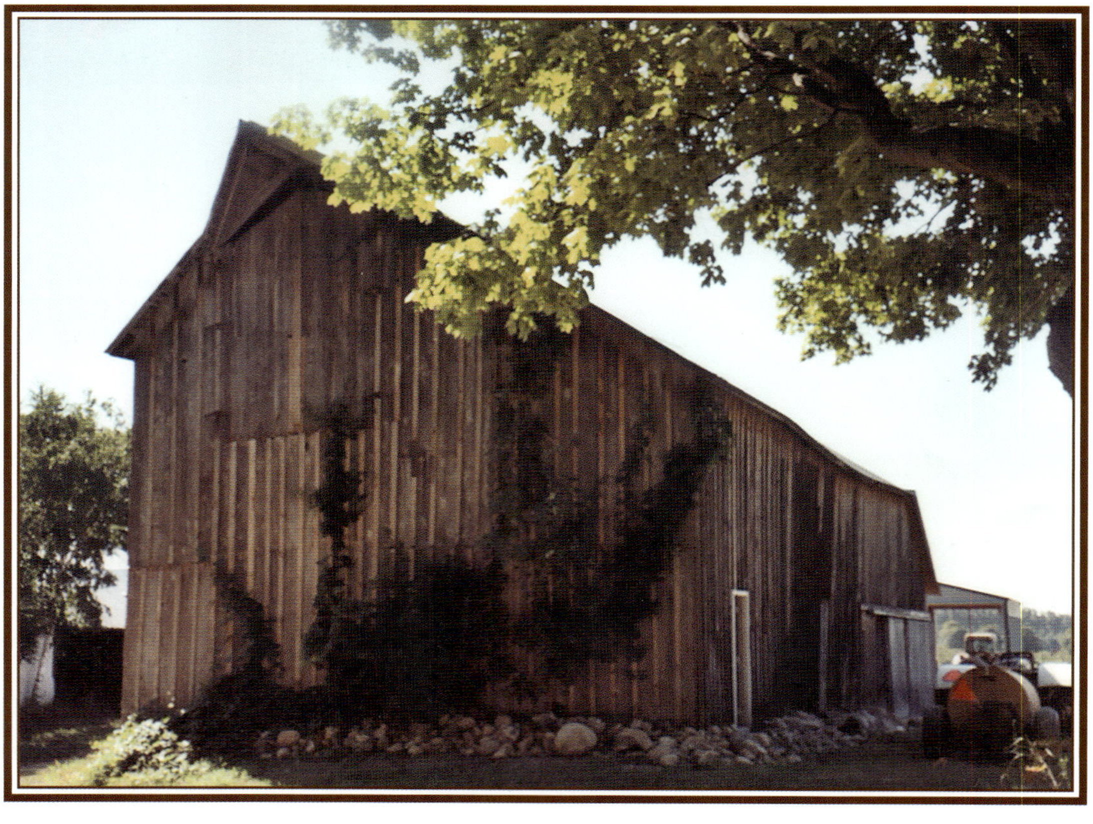

MANIGOLD BARN

"Mr. Sam Walker, a banker from St. Johns, Michigan purchased a large tract of land along the Bay before 1895. He had great expectations of turning the land into an experimental fruit farm. But he had some unfortunate business experiences and was unable to develop his land as he had planned. He had to make a paying proposition out of his farm. So, he took boarders during the summer months, and one of these was an author, and he put Mr. Walker and his farm into a book entitled, *Delightful Dodd*. According to this novel, Mr. Walker was an intelligent, educated man, with a sense of humor and a fondness of teasing, who had all manner of exciting things happen on his farm. Among other things the story included the capture of a bunch of desperados who had robbed the Elk Rapids Bank and pillaged numerous settlements on the Bay, using a stolen sailboat for their getaway (from E. Potter's book, *The Story of Old Mission*, p. 158).

The original barn was built by Mr. Walker in 1890. He lived there in one of the three houses on the property with his two daughters, Susie and Eloise. Eloise became one of the first woman doctors in Michigan. Ken Manigold Sr. bought the farm from the Walkers and farmed the original 120 acres. Likewise, Ken Jr. was a lifetime cherry grower and his son, Rob, following in his footsteps, now owns and farms the land. The original barn burned in the early 1940's and the present day barn was built. It is a large 30 x 60 foot timber frame barn built on grade with a big three wall additition and a hay hood on the gable end. It has a metal roof and a fieldstone foundation. This barn is board and batten on the outside and is unusual as one side of the barn across the gable end is longer than the other.

> ### Physical Description
> Date Built:
> - 1940's
>
> Construction:
> - Timber frame barn
> - Vertical board and batten siding
> - Off center metal gable roof with hay hood
> - Stone and cement foundation

The barn housed cattle, horses, pigs, and other animals. They grew cherries, apples, and prunes. Rob Manigold farms about 100 acres of cherries and also has about 10 acres of grapes for wine. Migrant workers were utilized before cherry shakers came into being. Rob told me that during cherry season they had 50 or so pickers on the farm. This would be in the 50's and 60's. They would generally pick about 500 lugs of cherries a day. Now that cherry shakers are used, the Manigold family alone can shake 60 to 75 tanks of cherries a day. One cherry tank tolds 40 lugs. Rob pointed out that this is done by three people.

The barn now is used mainly for storage. The Manigold Farm is recognized by the State of Michigan for its improved environmental practices through the Agricultural Environmental Assurance Program.

Ken Manigold remembers the Walker sisters, Susie and Eloise, being quite old when he was growing up on the farm. The main house, where the sisters lived, burned when Ken was 10-12 years old. He remembers the sisters were badly burned.

Information provided by Rob and Ken Manigold.

Writings by migrants inside a building near the barn

Summer

Winter

Unusual corn crib

Spider web

116

SNYDER BARN

This barn was sure to be demolished if Mr. and Mrs. Gene Snyder had not come along and bought it. It has been beautifully preserved and graces the hillside on Center Road, standing out from other houses and buildings because of its graceful lines and the meticulous care it has received. But first let's talk about a bit of history of this barn.

This barn has always been known in the neighborhood as the Boprey Barn. It was built in the 1920's and is 32 x 50 feet. It is a typical two-level barn with a gambrel roof. Its flared eave was a trademark of barns built by Ed Boprey, according to Cal Jamieson. It is built into the side of a hill and has a lower level on the road side. It is a double ramp barn, meaning you could pull a wagon all the way through the barn when needed. It has an unusual round corn crib sitting up on the hill on the side of the barn. Another unusual feature of the barn is that it is sitting perpendicular to the slope of the hill. Ed Boprey along with Bob Seaberg were known as local barn builders. They built this barn, and were known for building sawn timber barns.

Physical Description
Date Built:
- 1920's

Construction:
- Sawn timber bank barn
- Vertical wood siding
- Metal gambrel roof
- Poured concrete foundation

I talked to Barb Boprey, Bob Boprey's wife, who lived there from 1950-69. The house was on the other side of Center Road. They had 20 acres and raised cattle, hay, and grew cherries, plums, apples, pears, and prunes. She said her husband farmed, but since he could not make a living farming, he also worked at the Iron Works in town. They had 6 children, Barb says, and when she could, she worked at the Kroupa Cherry Station on Center Road. This was the maraschino cherry processing plant, but tart cherries were processed there also.

Her son always had a riding horse stabled in the barn. One day she heard a screech of tires and saw a car in the ditch. Her son's horse had gotten loose and the car had hit the horse. While the car sustained some damage, the horse was fine. Another exciting time was when a load of hay caught fire and, before they could put it out, their truck burned.

Cal Jamieson remembered that Ed Boprey always had a watermelon storage shed near the road (Center Road). Mr. Boprey stored apples there too. From time to time Cal and friends would snitch a couple of watermelons from the shed. One day, Mr. Boprey found the shed tipped over, and guess who got blamed? But Cal claims to this day he didn't do it. Such is life in a small community. Everyone knows, or thinks they know, what you are doing!

Bob and Barb Boprey sold the farm to the Mohrhardts when they moved to town. I do not know if there were other owners, but Gene and Jean Snyder bought the barn and a small parcel of land in 1987. It took three months for them just to clean out the barn. It seems the main floor had collapsed and fallen into the basement. Part of the metal roofing had blown off and was irreplaceable because of the unique pattern stamped into it to resemble shingles. Luckily most of the original roofing was found on the property and replaced. In place of he two sections the Snyders could not find, they cleverly installed sky lights.

"The front poured-cement supporting wall that faces east was falling away from the barn and had to be winched back approximately eight inches. Its calculated weight was approximately 20 tons," Gene Snyder told me. "All this had to be done before we could begin the work on the barn to try and put it back to original condition. We replaced the main floor joists with round poles as originally was used, and used 2 x 12 floor boards out of rough sawn boards. We did the same in the hay lofts….one loft is still an original. We built a new barn door for the basement which is facing Center Road. We rebuilt all four main floor barn doors, also replaced about five feet of siding on the north and south sides of the barn. All of the window glass had to be replaced as well as some of the window frames. The main fifty-foot pole that supports the barn is the original. All of this work was performed by my wife and myself, and we are very happy with the results. We hope it'll last another 100 years, so others can enjoy it also!" Gene and his wife Jean should be applauded for the restoration of one more barn on Old Mission Peninsula.

Information provided by Cal Jamieson, Barb Boprey, and Gene Snyder.

Old spray rig

SUNDEEN/EDMONDSON BARN

Ole Sundeen built this barn in 1927 but his name appears on the 1895 plat map with an earlier photo of the farm (1886) showing a similar dimension barn in the same location on the 1895 plat map. The barn is 32 x 32 feet and is a bank barn, so called because it is built into the bank. It has a fieldstone foundation on the back side. At one time it housed cattle in the lower level and stored hay in the upper level and loft. It has a good asphalt shingle gambrel roof. The siding is board and batten on one side and vertical shiplap on the other three sides, the shiplap being the oldest siding used. It has recently been painted a nice bright red.

Harold Edmondson married Elsie Sundeen in 1948, and helped farm the property. As with most of the farms on the Peninsula, the farm changed over to growing cherries in the 1920's. Harold told me that the barn was remodeled to house migrant pickers in cherry season. There were two bedrooms on each side of the three-bay barn. The center bay was made into a kitchen. Migrants were employed until the 1970's when technology in the form of the cherry shaker took over.

Dave Edmondson, son of Harold Edmondson, who now farms the acreage, told me that they have started reverting back to hand picking of the cherries. The cherries last longer when they are handled more gently. Cherries, in the past, were clipped from the trees with scissors to allow a longer time for the cherries to stay firm and fresh. He explained that leaving the stem attached allows the cherries to breathe and not deteriorate so quickly. They retain their juices and can be shipped great distances. Dave also said that they have been doing U-pick cherries (pick them yourself) for the public since about 1998.

Dave said the farm also uses contour plowing and tree planting. That means the rows of trees run perpendicular to the terrain, which is quite hilly. In between the rows of trees are sod buffer areas. This form of contour planting is used for safety reasons at harvest time and causes minimal erosion of the soil.

Information provided by David Edmondson.

Physical Description
Date Built:
- 1927

Construction:
- Timber frame bank barn
- Part board and batten and part shiplap siding
- Asphalt shingle, gambrel roof
- Fieldstone and cement foundation

PHANTOM BARNS OF OLD MISSION PENINSULA

Timber framing is characterized by the use of large dimension timber vertical posts and horizontal beams using pinned mortice and tenon joinery. This ancient method of framing buildings has been practiced all over the world for thousands of years. In 19th century (Michigan) American barn building, timber framing grew to its simplest, most economical and efficient form. As we enjoy a seemingly endless variety of shapes and sizes of (Michigan) American barns, the underlying timber frame is the highly efficient and elegant result of centuries of building evolution.

By late in the 19th century, American barn builders began to employ new technologies and experiment with new "modern" (and less substantial) building methods. Local steam-powered sawmills could produce an abundance of 2" thick planks, and the common use of wire nails helped lead builders to adopt lighter framing techniques. If well designed and well built with good materials, these "plank framed" barns have lasted nearly as well as timber frames. Because the use of the barns remained the same, i.e., diversified, animal powered farming, the massing and exterior forms remained the same through the change to plank framing. It is nearly impossible to tell from the roadside whether a particular barn is "timber" or "plank" framed. Once inside, however, it is readily evident.

Photo courtesy of Rex Shugart of the Ruby Ellen Farm Foundation

*Information on this page provided by Steve Stier,
Research Associate, Michigan State University, specializing in old barn construction.*

O.J. BENSON BARN

O.J. Benson came to the area in 1883, and was a farmer, fruit grower, contractor, and carpenter. However, what long-time Peninsula residents remember best about him is that he was a builder of barns. Whitney Lyon remembers he built "an awful lot of barns on the Peninsula." He also built the original St. Joseph Catholic Church and the Swedish Church that has long since been turned into a residence. His house, remodeled considerably, is now the Christie house on Peninsula Drive. The barn in the picture was taken down and moved down the road to the Edmondson's farm when the farm of 30-40 acres was sold off to make room for new homes. This old barn is now encased in metal siding to preserve it.

O.J. worked with his brother B.A. Benson who followed him to the Peninsula. Their barns were mainly gable roof, timber frame barns as the lumber was available on some part of every farm. Whitney Lyon estimates that the barns built by the Bensons were mainly built between 1905 and 1910.

Little is remembered about O.J.'s family, but he did have four children and all were educators. Walford taught at the university level. Two girls, Thelma and Esther, were teachers in the Detroit area, and another daughter was a librarian. They all would come and stay at the family residence during the summer. Ellen Sunquist Gray, Barbara Gray Springer's grandmother, told of having lovely tea parties with the Benson girls in the summer. They were real gentile ladies but they never married, thus there were no heirs to carry on the Benson name or heritage. O.J.'s legacy is in the many barns he built. Many are still standing and are noted in the individual stories about the barns. O.J. was in his late 80's when he died around 1940.

Recollections by Whitney Lyon and Barb Springer. Picture from Joan Bonney.

Physical Description
Date Built:
- c.a. 1890-1900

Construction:
- Timber frame bank barn
- Vertical wood siding
- Shingled gable roof
- Fieldstone foundation

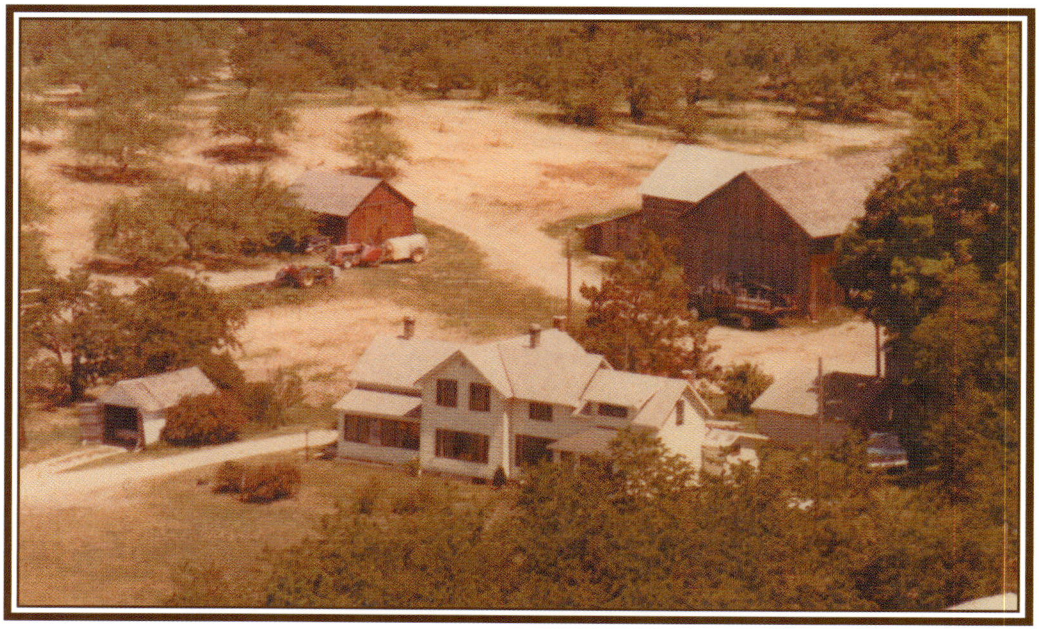

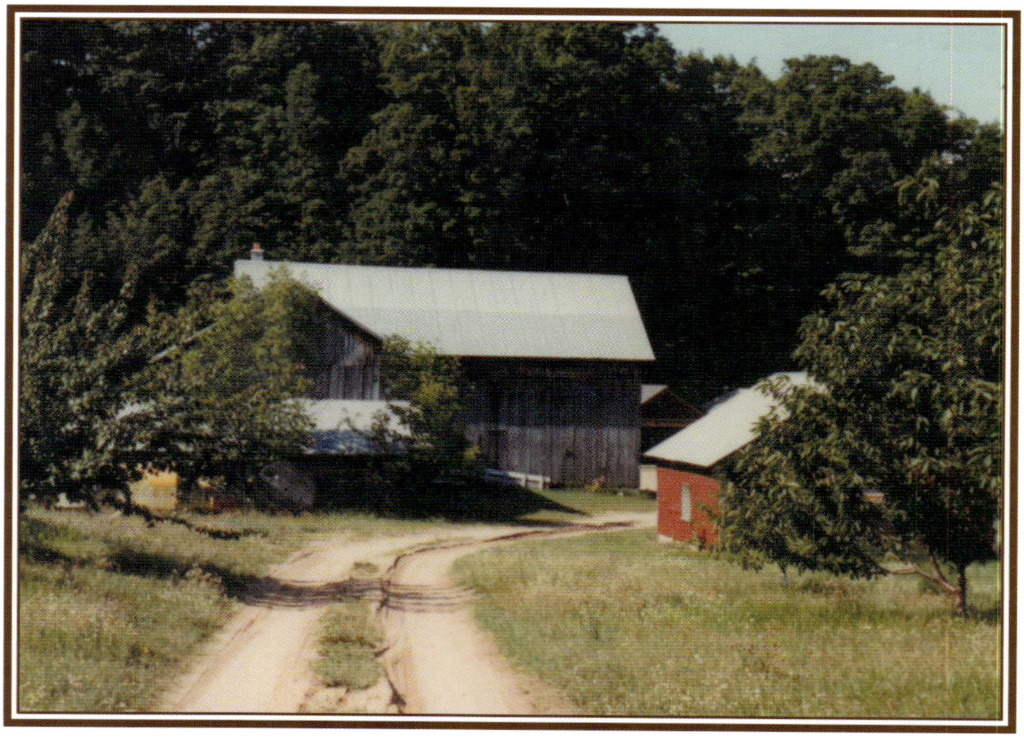

BUCHAN FARM

Five generations of Buchans have lived on this farm. I talked to Norm Buchan about the family farm and the barn that is no more. The Buchans are of Scotch/Irish descent and came from Canada in 1860. At one time they owned a large parcel of land, probably a quarter section (160 acres). Starting with William E. Buchan, the order of family living on the farm are: William H. Frank, Lester, and now, Norm and his wife Karen.

> *Physical Description*
> Date Built:
> - Before 1900
>
> Construction:
> - Early part, timber frame
> - Later part, sawn timber
> - L shape
> - Vertical wood siding
> - Metal gable roof on both parts
> - Foundation: unknown

The first part of the barn was in existence before 1900 and was built by O.J. Benson, a local barn builder. It was torn down and moved down the hill where it was rebuilt and added on to about the turn of the century. This created an L-shaped barn with a gable roof on both sections. It held the livestock (cows, horses, and chickens) and hay. The earlier part of the barn was timber frame and held together with wooden pegs. It was never painted. The added L-section was of sawed timber and bolted together. Frank and his son Lester farmed 106 acres of cherries, apples, peaches, and hay. Norm, the present owner, grew up with his six sisters and one brother on the farm. The farm was used for general farming to "feed the family" until 1955.

The farm house was also up the hill and later moved, intact, down the icy road in the winter. Logs were placed on the low spots and it was slid down the hill. Karen Buchan laughingly said it either slid or was held back, depending on how fast it was going. The house and barn were moved to be closer to the new West Bay Road. Before 1900, there was just a trail going along the bay. After the house was moved, it was gradually added on to and became very large. It was built like a duplex with two sides that were identical. At one time there were three generations living on the farm at the same time. Norm said, "It wasn't until 1948 that electricity was added to the house. It was available but Granddad Frank did not want to see the wires going across his property, so he would not allow Consumers' Power to hook them up. Also, Granddad liked farming with horses so they did not get a tractor until after 1948."

Some of the acreage down in the hollow was a wetlands area. It was considered a lake until it was drained. In 1939, Norm's dad Lester bought 6 blueberry bushes and planted them in this boggy area. Then he kept adding to them until they had five acres of blueberries. The blueberries thrived, and Buchan blueberries are well known in the Traverse City area in the summer. The farm is now just 15 acres with peaches and apples grown along with the blueberries. The bulk of the farm land was sold and the farm is nearly surrounded by subdivisions now.

Norm remembers that, as a kid, he and his sisters and brother had many chores around the farm…haying, milking, etc. Sometimes they went to town on Saturday. Cream was taken along to the dairy in five gallon jugs and was sold in order to buy the extra things that they could not grow on the farm. Norm remembers they always had a nice garden in the summer. He also remembers that they always went to baseball games to see his father play ball.

Norm bought the farm in 1975 and became a full time fruit grower. However, that job was done after he worked eight hours at a job in town. Karen said Norm loved to work with machinery, particularly cars, and used the barn in later years to repair and "tinker" with machines. The barn burned in 1988 after the Buchans had sold it for developing. Norm said it was in excellent condition and it was a shame to see it burn.

The Buchans' blue house sits high above West Bay overlooking Power and Bassett Islands, with splendid sunsets, providing we have sun that day.

Information provided by Norm and Karen Buchan.

December 30, 1940
Voted by the township board that the clerk would write to the State Highway Department asking for an amber light at the intersection of M-37 and U.S. 31 (on Front Street)

Colleen and Michele

The Alex Carroll Homestead

It seems the property the small barn is on was sold off at one time. (Tim Carrroll has since bought it back.) The owners at the time, Ken and Terri Pickett, were in the process of cleaning out the barn when Grandma Daisy Carroll wandered over from the house to see what they were doing. She was up in years, as they say, but had a strong voice. "Save the spiders!" she called out to Ken and Terri. She repeated it several times, and the couple, thinking probably Grandma was having some trouble with reality, promised they would. However they were puzzled as to how they would catch and keep those little, black, eight-legged arachnids when they found them. Later, they learned that the "spiders" Grandma was referring to were "black cast-iron frying pans originally made with short feet to stand among coals on the hearth." (Taken from *Websters' Dictionary*) Ken said that neither he nor Terri found any spiders for Grandma Carroll in the barn.

CARROLL BARN

In 1854 Richard Johnson met and married Mary Kaziah Rice in Traverse City, the eighth recorded marriage in the county. By 1864, the couple had accumulated enough capital to receive a deed to 160 acres on Old Mission Peninsula. The deed for the acreage was signed by Abraham Lincoln. Richard had come from New York as an Irish immigrant but received his citizenship upon moving to Traverse City. His wife Mary Kaziah had fled from Beaver Island with her family when King Strang took over the island. They named their new farm Sunny Slope as the house was built up on a rise near the road, with the barn being built down below and behind the house. The house, built in 1866, is still standing and looking even better than it did in the early photos. The current owner, Timothy Carroll, the fourth generation family member to live on the site, has restored the house.

The barn is no longer standing and few pictures remain, but we have enough to show what early life was like. The barn was built in 1870 by the family, with Richard Johnson's two sons and his brother Thomas helping build the gable roof bank barn. It was built down the hill from the house. The ramp and double doors were on the front of the barn which faced the road. It had two hay mows, one on the north end and one on the south end. The foundation was a crude sort of cement using sand with gravel mixed in, presumably from the shore. The barn had a shed extension on back where the hogs were kept. The timber for this timber frame barn came off the land. Tim has used some of the original barn beams in his kitchen as accents. You can see the marks made on them by the ax and the timber shows the mortise and tenon type of construction. It was always a red barn, but Tim does not remember ever seeing it being painted. The lower level housed cows and steer. There were eight stalls for the cows and several box stalls for the horses. Forty acres of the 160 acre farm were fenced for pasture. It was good for grazing but utterly worthless for growing anything else, Tim said. However the cash crop was always cherries. The farm needed over 100 cherry pickers when the crop was ready to harvest. The season could last up to six weeks.

Physical Description
Date Built:
- 1870

Construction:
- Timber frame ramped barn with large shed extension at back
- Vertical wood siding
- Shingled gable roof
- Fieldstone and cement foundation

When Mary Kaziah Johnson died in 1912, her youngest child, daughter Daisy, inherited the farm. She had married her childhood sweetheart, Alex Carroll, twelve years before and the couple moved back into her childhood house on Center Road. "It was a place full of fun," Tim remembers, "since grandmother Daisy played the piano and Alex played the fiddle and all the children danced and sang. That was the Irish way."

Alex Carroll had seven brothers, all of whom had cherry farms between Carroll Road and Blue Water Road. (This road was called The Church Crossing Road for many years, as St. Joseph Catholic Church was on the corner of Center and this road.) It seems each of the boys made their grubstake by logging in the winter. Alex was the 6th of 10 children of Edward and Jane Holman Carroll. The Holmans, the Buchans, and the Carrolls were neighbors in the previous generation in Wellington County, Ontario, Canada. They all moved to the Peninsula in the 1850's. An only Carroll girl, Elizabeth, married a Buchan and one Carroll son, Edward, married a Holman so they became related by marriage through the years.

Tim's father, Fred, and five sisters grew up on the farm. Fred continued farming with his father, Alex. In 1939 they went into diversified specialized farming. Thus they created work on the farm for a labor force year 'round. The winter project was raising mink. Tim told me that his father and mother drove to Kalkaska to purchase a pregnant female mink on March 28, 1939. The day is memorable in the family because Tim was born the next day (March 29), possibly as a result of the long bumpy drive to Kalkaska and back. The mink produced nine kits soon after, and a business was born. However Tim laughingly recalls his father stating that he (Tim) cost much more than that mink. Five 100-foot-long sheds were built to house the mink as their numbers grew. The operation was very successful, especially between 1946 and 1950. The Carrolls harvested around 1000 pelts a year.

The barn came down in the 1960's and not even the foundation can be found. There are other buildings on the property, one being Tim's writing studio. There is the structural remains of a forty foot high windmill just behind the house. The spinning wheel and blades at the top of the structure are long gone and in its place is a lovely angel with heralding horn that turns in the wind. It was specially made to Tim's specifications in Haiti. Tim seems very happy to return to the farm after spending his career traveling the globe as a diplomat. However, it is difficult to find him at home, as he still has a bit of the wanderlust in him.

Oral and written information and photos provided by Tim Carroll.

Migrant workers standing in front of the barn - 1930's

Picker's quarters

Small utility barn, Edson tractor, and workers - 1930's

The barn today

BRAUER BARN

George and Ethel Hill Boling purchased the farm in 1921 from Joseph Manley for $7,000. Several other owners of the property early on were Patrick Carmody, followed in 1881 by Mr. McDonald, then from 1895 to 1911, Earl Adams, and in 1913 Frank Carver held title.

> *Physical Description*
> Date Built:
> - early 1920's
> Construction:
> - Unknown - possibly plank frame
> - Horizontal siding
> - Wood shingled gambrel roof
> - Foundation is stone and concrete

George Boling Sr. was the editor of the Calumet Times newspaper in Illinois, and he and his wife had often vacationed in Traverse City. When George retired, they bought the Peninsula farm of 120 acres, naming it Boling Hill. Ethel was a very interesting person and was involved in many area clubs and activities. She and George had two sons, Robert and George Jr. After George Sr.'s death, Ethel and her family operated the farm for many years. After Ethel Boling's death in 1967, her son George and his second wife Frankie took over the operation of the cherry farm.

George Jr. had been in TV and radio in New York City before coming to the farm. The couple became famous for their beautiful peonies, which were prominently grown in the front yard of their home on Peninsula Drive. Everyone scattered, it is said, when Frankie Boling got behind the wheel of her big black Cadillac and drove to town. She was a woman of small stature, and it appeared that the car was driving itself, so everyone gave her plenty of room! Her husband George died of a heart attack in the orchard in 1972, and Frankie continued to operate the farm in a reduced capacity until 1997. Mary Lynn Sondee, a close friend of Frankie Boling, said she gave away what was being grown in her last years on the farm. She died in 2003.

Richard and Marty Brauer bought the 76 acre property in 1997. They have renovated the original farmhouse into a comfortable cottage-like home and have planted hay in the former orchards to build nutrients in the soil with the anticipation of raising fruit on the property again. The old orchards were pulled out as they were too old to be productive.

There was an attractive three-level barn on the property at one time, close to Peninsula Drive. It had horizontal siding and a cedar shingle roof. It looks to be a plank frame barn and was built while George Boling Sr. lived on the property, in the early 1920's. Unfortunately this barn burned to the fieldstone foundation in 1961, injuring four. Many pickers had been housed in the barn. Today one small tractor barn still remains and two somewhat larger barns up in the hills behind the house which probably housed another 100 or more pickers at the peak of the manual cherry industry. One row building held many small "apartments" and every apartment door was painted a different color. A tree along Peninsula Drive still shows engravings from these seasonal workers.

The farm was included in the Peninsula Township's PDR (Purchase of Developmental Rights) program in 2004. The Brauers continue to find artifacts of years gone by, including toy metal cars discovered at the edge of the woods.

Information from Rich and Marty Brauer, and Mary Lynn Sondee.

> Now for a story with a real twist. Frank Carver left the Brauer farm and moved *to* 628 Washington Street in Traverse City to live around 1916 or so.
> When Rich and Marty Brauer bought the farm in 1997, they moved *from* 628 Washington Street in Traverse City. They lived there for 19 years and raised three kids: Grace, Gretchen, and Colin.

John Hemming Sr. farmstead

Twister Hits Area, Damage Is Heavy

A tornado spawned by severe thunderstorms slashed across Old Mission peninsula and Antrim county Friday night, smashing barns, at least one house, other buildings, and countless trees.

No deaths or injuries were reported, although the total path of destruction was about 30 miles long and up to a quarter of a mile wide.

Appearing about 10:00 p.m. out of a heavy thunder and wind storm, the twister sliced across the John Hemming farm at Bowers Harbor on Old Mission peninsula 11 miles north of Traverse City, crossed East Grand Traverse Bay, then cut another swath of destruction across Antrim county from north of Elk Rapids to Echo township in the north-central part of the county.

A barn and four worker cabins on the Hemming farm were flattened, a $1\frac{1}{2}$-ton truck was carried more than 15 yards and dumped upside down, and a spray rig laden with 500 gallons of water was spun around. Near the Hemming place, a new summer cottage believed belonging to Alex Misekow of Flint was demolished.

Officers said the Hemming house was not hit by the tornado, although the family had sought shelter in the basement, but that Hemming was spun around inside his house by the suction of the wind just as he reached the cellar door and that one of his children was slammed against a wall but not hurt.

Deputies said Hemming described the twister as sounding like numerous freight trains and said it arrived after an ominous lull in the roar of the thunder storm.

—*Traverse City Record Eagle*, May 9, 1964

JOHN SR. *and* VIRGINIA HEMMING BARN

This farm has a long and interesting history. "Captain Frederick Johnson was a native born in Mexico in 1829, but the great republic of the United States became his land." (*Sprague's History*, 1903, p. 765) He served under General Zachary Taylor in the Mexican American War.

"Johnson's first trip to the Grand Traverse region was in 1852 when he and a Mr. Whelpley surveyed and cut the first road along the head of the east bay between Traverse City and the Five Mile Corners. He visited the area several other times and then settled here permanently in 1856. On Christmas day in 1855, Frederick Johnson and Susanna Lother were married in Chicago. In 1856, the young couple obtained a homestead near Bowers Harbor on the peninsula and began to clear the land and make a home for their future family." (*Grand Traverse Legends*, p. 109) "At one time his land purchases had reached 640 acres. After he retired (in 1891) he owned about 400 acres with 240 being improved, and his property now constitutes a very valuable and productive farm." (*Sprague's History*, p. 765)

Physical Description
Date Built:
- 1939

Construction:
- Plank frame with large timber supports
- Vertical wood siding
- Asphalt gable roof
- Fieldstone foundation

Maurice DeGraw tells about his grandfather, Captain Johnson, in *Memories Hidden, Memories Found* (p. 207-8) "The farmland up in the hills on Neah-ta-wanta was considered to be good farmland. Captain Johnson brought his bride to the area, and built a cabin, and much later, a beautiful big house. Captain Johnson was a sailor and a captain on several boats. He was gone from early spring and all during the summer. He would come in on Saturday night and his boat would lay up over at Elk Rapids Harbor. He would take a rowboat and row across the bay to Old Mission Harbor, and walk across the Peninsula to his house on Neah-ta-wanta. Sunday night he'd walk back over to Old Mission, and row back across to Elk Rapids. You see, he sailed only on the inland lakes at this time, and there were rapids at Elk Rapids that he could not negotiate with the ship. The inland lakes boat was a flat bottom, a type of river boat. It was a passenger ship that he ran for twenty or thirty years. My grandmother did the farming with the help of the five boys in the family. There were five girls too. They had beautiful big barns."

Maurice continues, "Captain Johnson was well known (and respected in the community). He loved to entertain especially in the winter months. Grandmother was busy most of the time having parties for people from Traverse City, and out here, too. He always had a fancy team of horses they kept just for him. He'd smile and nod his head to everyone along the road."

The barn that is in the picture was built in 1939 by the Corells, who had cattle and grew cherries on the acreage from 1939 to 1955. John and Virginia Hemming bought the farm in 1955, increasing the acreage to 240 acres. They found an old foundation showing where a barn was previously. The "new" barn was a gable roof bank barn. The Hemmings raised harness horses…trotters and pacers. Two mares were in foal when disaster struck. May 8, 1964, a tornado tore in off the harbor and twisted across Hemming's farm. The barn came down on the horses stabled on the main floor, and barn siding boards were strewn all over the area. The twister also flattened four worker's cabins. A summer cottage on the shore, in front of the farm, was demolished, and a picture shows the trunk of a good sized tree driven into the roof. A one-and-a-half ton truck was moved more than 15 yards and turned upside down, bringing down the electricity wires on top of it. A spray rig loaded with 500 gallons of water was turned, facing another direction, when found. Shingles from the barn roof were driven into trees and dead fish were found up in the yard. John said a ten foot deep cavern was washed into the hillside from the force of the wind and water. It is no surprise that the *Record Eagle*, May 9, 1964, carried a front page story and several pictures of the damage. Luckily the lovely homestead, just west of the barn, was not hit.

The barn was not rebuilt. A roof was put on the remaining foundation and is used for storage. John Hemming was in the plumbing business, Walters and Hemming, for many years. John says that his wife Virginia ran the farm and was a very hard worker, while he worked in town.

In 2005, John and Virginia Hemming moved from the farm, after living there 48 years. Their son Jed and his wife Dawn, who have owned the property since 1983, are in the process of renovating the house, including moving it 60 feet to the east. It has a new foundation and a better view of Bowers Harbor.

Information and pictures provided by John and Virginia Hemming.

L to R: John Hoffman by his oxen, Aunt Lizzie Hoffman Curtis, Grandma Matilda H., Baby Aunt Lillian H. Carroll. The rest are unknown.

William Hoffman with son Tom and friend in front of the barn

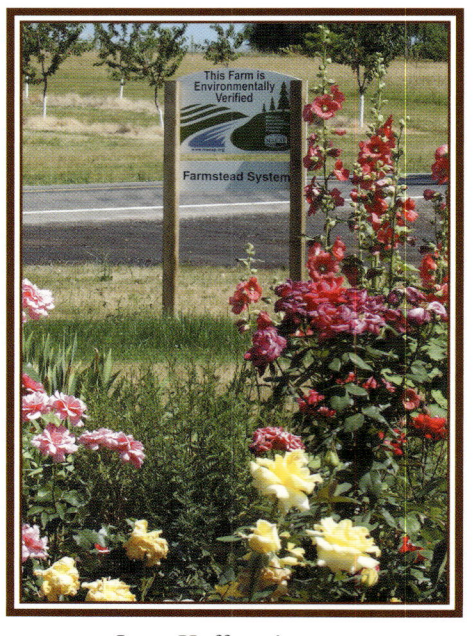

Irene Hoffman's roses

TOM and IRENE HOFFMAN BARN

The Hoffman Barn is the barn that was! When the barn was no longer used, Tom Hoffman, third generation on the farm, had it taken down in 1978 and it was used as décor for The Bowery Restaurant on Old Mission Peninsula. However, this centennial farm has been in the Hoffman family over 125 years. The barn was moved back from its original location in 1931-32. Tom explained that a track was built to move the barn. Then the barn was put on rollers with timber underneath it. A horse would go around and around, winding a cable attached to one of the barn's cross timbers, moving the barn forward a few feet at a time. The barn was moved by means of a large winch, powered by a team of horses. A rugged fieldstone basement was added to the barn later. The stone foundation was built by Steve Steith, and still remains.

I saw the land grant of 160 acres that were originally granted to the Perkins family to reward Mr. Perkins for time served in the Civil War. However, after he died and Mrs. Perkins remarried, it belonged to the Fowler family. Tom's grandfather, John Hoffman, came from Grand Rapids with a team of oxen. He married Matilda Lardie, a local girl. In 1879 they purchased 80 acres from the Fowlers for $1,400. The barn was built first, most likely in 1879. The oxen became a great asset for John and Matilda Hoffman. Tom Hoffman said, "Granddad would take his oxen to Joe Kroupa's farm and work one whole day pulling stumps, hauling logs, and other heavy duty chores. Then Joe would come down and work three days helping out the Hoffmans." Tom believes that a good deal of the 80 acres was logged off by the time his grandparents bought the property. Thus they could begin planting apples, cherries, and potatoes. They also had cattle.

Physical Description
Date Built:
- First barn 1879
- Another barn at a later date

Construction:
- Timber frame barn
- Vertical wood siding
- Wood shingle gable on one side, metal on one side, gable roof
- Fieldstone foundation

They built a house of rough sawn basswood, and attached another building to the house, which was a grocery store for the area folks. Later, a lean-to was added to the house/store. Tom's father, William, was born in the house in 1885. In 1892 a 'new' house was built using cedar shake shingles and sawn lumber from the farm. A part of the house Tom and Irene Hoffman now live in is from the original building. The rest of the store/house was torn down in 1938. The same year, Tom's father was working in the barn when he was gored and trampled by the bull. The "tame" bull was supposed to be in his own pen, but the terrible accident happened and William was killed.

Tom was always a farmer, retiring in 1984. In earlier days, he remembers working out to earn extra money for his growing family. He had marrried Irene in 1948 and they produced seven children. One of his outside jobs involved working at the livestock auctions at the old fairgrounds once a week. He also hauled ice from town to fill ice houses around the Peninsula. Another job was fleshing out mink (skinning them) for Oakley Lardie and Fred Carroll. He carved all the pegs that went into the Ray Dohm barn, just across the road.

Son Bill now owns and farms the land, with cherries and pumpkins being his cash crops. In the fall, the Hoffmans' yard is a sea of orange as Bill raises a lot of pumpkins. It is a family affair as they gather to help pick the pumpkins and sell from their yard on Center Road. Today, we think of Tom and Irene as the folks that always have a splendid garden in the summer.

Information and pictures provided by Tom and Irene Hoffman.

Curtis and Louisa Fowler's first home

FOWLER/HOLMAN BARNS

The Fowlers and the Holmans were among the earliest settlers on the Old Mission Peninsula. I talked with Jack and Georgia Holman. Jack Holman can claim close family ties to both families. Great-great grandfather Curtis Fowler, Sr. came from Vermont. He and his wife Louisa first settled in Jackson, Michigan. In 1855 Curtis and his two sons, Curtis Jr. and Francis Z., followed the Indian Trail from Grand Rapids to the Peninsula. They each bought 160 acres of land. In 1856 Curtis Sr. brought the rest of the family to the Peninsula. The family lived in a log cabin on Center Road. He was an important figure in the community—a Probate Judge of Grand Traverse County for 14 years and the postmaster at Mapleton. But primarily he and his sons were farmers, and they started growing apples in 1861.

Both of the Fowler boys served in the Civil War. Curtis Jr. enlisted first. He was wounded in April 1861 and returned home to farm. His brother, Francis Z. Fowler, also went to war and was killed in the Battle of Manassas in 1862. The land belonging to Francis Z. Fowler was on Center Road and was a part of his father's land. This land was later purchased by John Hoffman and is still in the Hoffman family today. Curtis Jr.'s 160 acres were northeast of his father's land on East Bay.

> **Physical Description**
> Date Built:
> - c.a. 1900
>
> Construction:
> - Timber frame on grade barn
> - Vertical wood siding
> - Original cedar shake gambrel roof covered with asphalt shingles
> - Foundation: unknown

On the maternal side of Jack Holman's family, his ancestors came from Elora, Wellington County, Canada. Alex Holman came to the Peninsula to visit his Aunt Jane Eliza (Holman) Carroll, who was the wife of another early settler, Edward Carroll. Alex met Ethelwyn Fowler. They eventually married and took 40 acres on Center Road, part of Ethelwyn's father's 160-acre farm, Ethelwyn being Jack Holman's grandmother. This farm is now a centennial farm. The Buchan, Carroll, and Holman ancestors came from Canada about the same time in the later part of the 19th century. They lived near each other in Canada, and the descendants of the three families are still living on the Peninsula.

George McManus, Jr. tells a story about Jane Holman Carroll to show a piece of pioneer life in the 1800's. When Edward Carroll came to the Peninsula in 1863, his wife Jane stayed behind in Canada until Edward had secured a 160-acre tract of land. As soon as a log house was built, Edward sent for his wife. She came by train to Traverse City, and then she walked the eight miles out the Peninsula to the farm, with two young children. One was being carried on her back, and another young child was led by the hand. A short distance from town, a bear came down out of the woods and crossed the road in front of her. More than a little shaken, she continued on until she came to a farm on Carpenter Hill. The owner was milking his cow. She asked for help. "Well," he reportedly said, "when I finish milking, I'll hitch the cow to a cart and take you up the road to your place." And that he did. George McManus said he heard his father tell that story many times.

Jack Holman told of growing up on the farm on Center Road with his parents, Bernard and Ruth. When Jack was a child of nine, he said, he drove the tractor they had bought in 1937 while his dad sprayed the fruit trees. He said that first tractor was no bigger than a good size riding lawn mower of today. Before that, Jack said they had a team of horses, named Jack and Nell. The Holmans always had several cows, hogs, and chickens. They started growing cherries very early. Jack remembers being told his Uncle Frank started an orchard in the late 1890's. He thinks they were sweet cherries and says he was told of a two acre block of cherries up on the hill behind the barn. The Holmans also had a two acre peach tree nursery orchard and sold peach trees to Hawley's Nursery in Hart. Jack remembers an Indian trail ran along Bluff Road and the Indians had summer gardens up on the high ground between Bluff and Center Roads. Their trail would have been about where Cherry Hill Road runs down to Center Road today.

In the winter of 1932, the original barn on Center Road burned down when a lantern tipped over. Tom Hoffman remembers watching the fire from an upstairs window in their home across the road. His folks wouldn't let him go help put out the fire because he was too young. Neighbors packed snow around a nearby shed to save that. Some of the older boys threw snow balls at the fire to "help put it out." Tom also remembers that later that year the neighbors gathered for a barn raising "bee." The men brought their hammers and saws, and the women brought plenty of food. A new barn was up in no time at all.

Jack recalls that a water supply for the neighbor's oxen came from a flowing spring next to Center Road on the Holman property. John Hoffman sunk a wooden barrel into the ground around the bubbling spring, and the barrel was always full of fresh water. A sprout of a willow tree grew off that barrel. (It had a rim made of willow.) The tree grew to be very large and had a split trunk. Jack said his sister and he would climb into the V of the tree to wait for the school bus. When that tree was taken down a few years ago, several sprouts came up. They were all cut away, but one managed to survive, and if you drive past the Tim Holman farm today, you will see a nice sized willow tree next to Center Road.

After Jack married and returned from the Army, his parents retired to a home on Bluff Road. Jack and his wife, Georgia, moved into the home on Center Road where they lived and raised their two children. They continued to live there until about 1997 when they retired to a Bluff Road farmstead of 56 acres they had purchased in 1967. Son Tim and wife Laurie, an avid gardener, moved into the Center Road home and have taken over the farming operation. He farms what is called a "long forty," which means the property is much longer than it is wide.

The farm property Jack and Georgia purchased from his father's cousin, Zeb Fowler, on Bluff Road had a barn that was most likely built at the turn of the century. The floor of this unusual-shaped barn was earthen and wood. It had a loft, but no lower level. Because of the low roof, it could not be used to store large equipment. It sat unused until Jack gently pulled it down. He said it didn't take much pulling for the barn to collapse, since it was so badly deteriorated. The roof was the original cedar shakes, but it had been covered with tar paper and asphalt shingles. At one time the barn was home to cattle and other livestock, with hay in the mow up above.

The willow tree today

The 1903 farm house on the site offers a step back in time as soon as you enter the back door. Everything remains the same as when it was built, from the original wood floors to the wide door casings, and the many period pieces of furniture that grace this lovely old house. The house, which has been added on to, was moved around 1915 from a location just north of where it is now.

The Holman's grandson, Cory, now a teen-ager, claims he will be a farmer someday too. Their son Tim is a sixth generation fruit grower on the Peninsula, making Cory a possible seventh generation fruit grower!

Information and pictures provided by Jack and Georgia Holman.

Early peach tree orchard

Largest Tree Fell – 1904
What was probably the largest tree on the Peninsula was cut and sold recently to the Wells-Higman Co. Six twelve-foot body logs, scaling 7,651 feet and 4 logs from the top, scaling 704 feet, or a total of 8,305 feet, were cut from this one elm tree. $113.55 was netted by the seller, Mr. Holman.

HORTON BARN

This simple gable roof barn was torn down in 1970. It was probably a turn of the twentieth century barn. The picture, taken in 1947, shows Arlene Horton and her daughter in front of a noticeably sagging barn. The Hortons purchased the farm from Mae Cowan. While the Horton family never lived on the property the barn was on, they owned the 160 acres that included the barn. Arlene told me the barn had housed cattle and had a loft for hay but did not have a lower level. Another cement block building sits on the foundation today.

The significance of this barn and land is that it was the first farm in the state of Michigan to be sold to the American Farmland Trust. Working with the Township Zoning Board of Appeals, the 160 acres could not be developed. Four parcels of land that the Hortons owned could be sold for developing. Restrictions were imposed as to what could or could not be done with the land. The Hortons sold it in 1992. As Jim Horton said at the time, "As long as it is in farming, I feel good about it." The Hortons lived on the property a short way up Tompkins Road until 1997. They had retained five acres and the house. After Jim Horton Sr. died, Arlene moved to town.

Physical Description
Date Built:
- c.a. 1900

Construction:
- Timber frame barn
- Vertical wood siding
- Metal gable roof
- Foundation: unknown

Among Arlene Horton's memories is that she was not just a farm wife doing "woman's" work in the house. She remembers picking and hauling cherries to the Peninsula Fruit Exchange. She said, "Jim was a farmer from day one." He loved the farm and really loved his animals, horses and dogs the most. Their two children had riding horses when they were growing up. The Hortons were cherry growers for fifty plus years on their 160 acres. Arlene said that the migrant workers that came to work for them were primarily Hispanic, coming from Texas and Florida. Many of the families became good friends. One worker in particular stands out in her mind. One time when Jim Horton needed to be hospitalized, Señor Martinez came to the hospital to visit him. That was special, Arlene said.

The former owner's daughter, Mae Cowen, lived on the acreage where the old barn was standing. Arlene remembers Mae, a registered nurse, was always helping people and animals with health "issues." She was lovingly called the Doctor of Old Mission. Arlene remembers one time when they had a sow about to deliver her litter and something was terribly wrong. It was in distress and could not seem to deliver them. Arlene's husband and hired man were out on the bay fishing and, while they managed to "yell" them in from the bay and deliver the baby pigs, the mother sow died. Mae took the piglets into her house and fed them…all eight of them…around the clock. They all survived, thanks to Mae's skills and dedication. Arlene said she was a special lady and kept all sorts of feathered friends in the barn. She had ducks, peacocks, turkeys, geese, and chickens. It was a sad day when she died in 1977, according to Arlene.

Information and picture provided by Arlene Horton.

The American Farmland Trust was the forerunner of the Peninsula's PDR, Purchase of Developmental Rights Program. The rules and restrictions are different, but the concept is much the same. The object is to preserve farmland and open space from being developed.

Ivan "Doc" Jamieson farm
on Smokey Hollow Road

Cal Jamieson barn
looking toward the bay

Cal and Verla Jamieson barn on Bluff Road

JAMIESON BARNS

This phantom barn was originally built and owned by Perry Fouch. Before Cal and Verla Jamieson bought it in 1954, it had two other owners. First Tom Simmons and then Joseph Tiffany owned the property. The farm with 40 acres of cherries, was just north of the barn and the acreage was expanded to 85 acres. They also had a 90-acre cherry orchard on Neah-ta-wanta Road. Cal farmed with his father and later his stepfather on another 90 acres on both sides of Smokey Hollow Road.

The barn itself was a three-story bank barn with a shingled gambrel roof. The front of the barn had an extension and another in front of that for added storage. Farming was never easy on the hilly terrain of the Peninsula. Verla remembers a few seasons in the 60's and 70's when she was "boss of the crew," meaning she was the overseer of a crew of 7-8 young men who manned the cherry shaker operation in the orchards. Cal was "working out" at the time, driving a double bottom tanker truck for Wagoner Transportation, hauling gas and oil all over Michigan.

Physical Description
Date Built:
- Early 1930's
Construction:
- Timber frame barn
- Vertical pine siding
- Metal gable roof
- Poured cement foundation

Verla remembers some harried times on the farm. One time when the ground was slick from the morning dew, the cherry rig literally slid down the hilly orchard with the tractor driver standing up on the brakes, to no avail. Another time, son Kim was trying to rewire a light in the barn, and he was having some difficulty making it work. The wiring started to smoke, sending white smoke billowing out the barn window. Verla said she knew the barn was on fire, one of the greatest fears of all farmers. Quick thinking helped save the barn that time. Verla was involved in other things besides farming too. She was active in Old Mission School activities and was a member of the school board at one time. Also, with her soft, but no nonsense voice, Verla was the moderator on an early morning T.V. show for a short while in the 1960's. The program, Accent Agriculture, was featured on Channels 7 & 4. It was about current events, anything that had to do with farming.

In 1964 a tornado came through the Jamieson's Neah-ta-wanta cherry orchard. The Hemming barn was directly in the tornado's path and was demolished. Cal's orchard was up the hill behind the barn. The force of the storm threw shingles and nails up into the trees. Cal said he and his son found nails for weeks scattered all over the orchard.

Eventually the family barn on Bluff Road was torn down because it was standing in the way of progress. The Jamiesons sold their farm in 1973, and the acreage was eventually developed into Mission Hills Subdivision. The Jamiesons now live in a comfortable home looking out on East Bay on one side, and on the other side they see the hillside, where their barn and orchards once were, now dotted with lovely big homes. But the . . memories remain.

Cal reminisced about life on the farm when his father, Ivan, "Doc" Jamieson, farmed over on Smokey Hollow Road. Cal said this barn was hauled from Northport in the 1940's. It was torn down in Northport and the huge timbers were rafted from Omena around to East Bay. The lumber for the siding was brought around the bay on a truck. The barn was then rebuilt on his parents' property. When the barn burned down somewhere between 1949 and 1951, a great deal of farm equipment was lost, including the tractor. No picture exists of the old barn, but an aerial view of the barn foundation and house are pictured.

Ivan Jamieson owned a large, beautiful stallion named Big Boy, who provided stud service. The small barn belonged to the horse. Big Boy visited many Peninsula farms, as well as Kalkaska County and Leelanau County farms. The first two summers Big Boy serviced more than 200 mares. He lived to be 14 years old. Doc Jamieson was known to be good with animals and helped other farmers when they had a sick animal. In fact, that is how he got his nickname, Doc. Cal remembers his father was also a good cherry farmer. Ahead of his time, Doc was the first to sod orchards to prevent erosion on the Peninsula's steep slopes. Now, all cherry orchards have grass growing in among the cherry trees. Also, Doc planted corn between the small cherry trees. He thought of it as wasted space, so the family made good use of the ground between the trees until they grew larger.

One more story Cal told me about growing up on the farm took place when he was young, about 10 years old. Some buddies and he decided to disassemble a barn wagon and reassemble it on top of the Mapleton School. They worked hard on this project for hours. When they came down off the roof, they were met by Cal's dad and the local sheriff. For all their jolly good fun, they had to bring the wagon back down, like they took it up, and reassemble it on the ground. Then they took it home!

Information and pictures provided by Cal and Verla Jamieson.

OCANAS FARMS

When Leo and Carmen Ocanas bought their 22-acre farm in 1978, it fulfilled a dream Leo had strived to achieve for many, many years. He had saved $29,000 to buy George Penny's run down farm. The farm had a house, a stone garage, a couple of out buildings, and an old barn. This story talks a little about a phantom barn, as Leo tore it down shortly after buying the farm. It is more about Leo, a Hispanic migrant worker from the south and his accomplishments and contribution to the Peninsula.

The weathered red barn had a shingled gable roof, with a metal roofed three-wall shed attached on the north end. The barn was on a nice field stone foundation that still stands. The inside of the barn had a loft. At one time it was a barn that held horses, cattle, hay, and small farm machinery. It probably was a "general purpose" barn for a "self-sufficient" farmer. It had simply outlived its usefulness. Leo had bigger ideas for the farm.

Leo Ocanas had a goal to grow the best orchards and has said he will never stop until he has accomplished that goal. He says he started without much equipment to farm and without much orchard. When he started out on the farm, Al Carroll loaned him the farm equipment he needed to tend the orchards. Then Leo would work for him in return. He did get his first cherry shaker the following year in 1980. He smiles when he says he still has the 1949 Ford Jubilee tractor that came with the farm. He basically worked alone, but Carmen helped when needed, running the harvester.

Physical Description
Date Built:
- Unknown

Construction:
- Probably timber frame with extension
- Vertical wood siding
- Gable roof
- High stone foundation

"Those first orchards grew to more than 350 acres, as Leo took on contract work for other farmers, including Nancy Heller, whose cherry trees Leo used to pick as a child." (*With These Hands*, p. 18) Leo said, "Farmers on Old Mission Peninsula work together. They look after each other."

Leo, his mother, and four older brothers and sisters were migrants that came up from Florida in the summer and fall to harvest cherries and apples in the Old Mission area. They followed the crops to other states during the year as well. That is how, working together, they made a living. "Life was hard: long hours, low pay, tedious labor. But Leo remembers it as a good life. No one went hungry, because everybody worked, and they always had what they needed. 'Just enough,' he says, 'just enough' (From *With These Hands*, p. 18). Leo said he started working alongside his family in the fields at the age of six, and never attended school after grade three.

One story Leo told me was about when they got their first new car, a 1960 Chevy Bellaire. The car dealer was not interested in helping them until his mother produced cash…$2,900! Leo drove his family away in their new car. He was 14 years old, the family's driver, driving on a restricted license. Although he was the youngest, he was the family's leader and was always a dynamic force. Being a Hispanic made him even more determined. "I was going to prove to myself that what you are really doesn't mean much," he says, "It's what you can do." (From *With These Hands*, p. 19)

Now Leo produces more apples than any other farmer on the Peninsula. However, he also grows peaches, apricots, nectarines, and Santa Rosa plums. Some of the other farmers told him the plums, which are grown in California, would not grow successfully in Northern Michigan. But Leo likes a challenge and he is raising fine crops of plums as well as the rest of the fruit. Leo said, "It bothers me when I see a dead tree in my orchard. I must analyze the problem and fix it. I will keep only healthy trees." He now farms 165 acres of his own on different farms and contracts out other farms for a total of 300 acres. On his northernmost farm Leo has 104 acres in the Purchase of Development Rights program. He does his own pruning and fertilizing but does have help from long time workers who have become good friends. They have come from the south for the last 20 or more years, arriving in April and staying until November. He told me he never gets tired of farming, but he produces according to his finances. Some years are better than others and, he lamented, the "pricing was awful."

"The best thing in my life was when I met Carmen," Leo said softly. "I was already 26 years old when I met and married Carmen in Florida." Working almost round the clock did not give him much time for romancing. Carmen, likewise, is a very hard worker as she tends a three-acre garden and takes the produce and fruit to four different markets in the area each week. She also keeps the record books for the farms. That alone is a big job, Leo admits. They have two grown children, and Leo doesn't think Leo Jr. or Angie will become fruit growers as they have other interests. The kids do help in the summer when they can with the orchards or the farm markets. The apple of the Ocanas' eye (pun intended) is Emily, their youngest grandchild.

"When the Ocanases came for the harvests 40 years ago from Florida, Leo and Carmen remember the Old Mission Peninsula covered with orchards as far as they could see up and down the roads." Not so anymore. "Some farmers retired or died and their children sold the land for less fruitful—but more profitable—uses. Leo says the land prices make it almost impossible for farmers to buy land—much less hold onto it. The only way he could afford to expand was to buy acreage from a land owner who lowered the asking price of his farm by first selling its development rights for preservation. "There's no two ways about it," Leo says, "we're going to see some houses and some condominiums in this area." (From *With These Hands*, p. 25)

Now, Leo and Carmen spend a number of weeks in Florida in the winter, warming up, slowing down, preparing for the busy season in the coming year. Life has come full circle for them!

Oral information and pictures provided by Leo and Carmen Ocanas.

Granddaughter Emily

The farmstead

PRATT/ALTENBURG/LEVIN BARN

Jerome Pratt was one of the earliest settlers to come to the Old Mission Peninsula. He came with wife Araminta whom he had married in Ypsilanti, Michigan, June 18, 1845. Jerome served as an Indian Agent for Reverend Peter Dougherty when he moved to the woods of the Peninsula in 1849-50. He ran the Sunday school, and spoke the Indian language very well. Araminta was popular with the native people also. The couple bought land from H.K. Cowles in 1854 and built a log cabin on it, and later framed it in. The house stood in front of the lovely home on the property now. Jerome was always a farmer and had 140 acres in Old Mission, with some of the property going down to the bay. He had added to his land in 1864 when he bought land from Robert Armor's widow. Her husband was given the property in appreciation for his service in the War of 1812. Both documents of the purchases were found by Molly Levin's son and daughter-in-law, Steve and Nikki Sobkowski. The barn was most likely built around 1900.

Jerome Pratt had an interesting career and was not just a farmer. He moved his family to Cross Village and became the lighthouse keeper at Skillagalee Point for a short time. Araminta taught in the first Indian school there. The Pratt family moved back to Old Mission, and Jerome became the first lighthouse keeper at Old Mission Point in September 1870 when it was completed. Jerome was the keeper for three years and, according to his original log, "the light was burning and shipping was open on Grand Traverse Bay from April 5 to December 20. From two to ten ships were sighted daily and most of them were schooners. There were no wrecks but there were several gales." (Dr. Potter's Book, *Story of Old Mission*, p. 70)

Physical Description
Date Built:
- c.a. 1900

Construction:
- Timber frame with side and back extensions
- Vertical wood siding
- Shingled gable roof
- Foundation: unknown

Molly's grandparents, Will and Mary Pratt, built the present house which is a lovely, large home in the Queen Anne style in 1902. Molly told me that her grandmother got the plans from a magazine. It is one of the few, if any, old houses of this style left on the Peninsula. The barn out back was of the same era. Much activity centered around this farm through the years. Molly's grandmother always had a huge garden, horses, cattle, chickens, and hogs. Grandmother Mary would make chicken dinners for the Old Mission school children, and for 10 cents they could walk across the road for their lunch. Teachers were boarded upstairs in one of the many bedrooms.

According to Molly, she spent a great portion of her life on the Peninsula on this farm. Molly's mother and father, George and Helen Altenburg, came every summer to the farm from Highland Park, Michigan, where Molly's father was dean of the junior college. Molly remembered there was always a lot to do in the summer working on the farm. There were 13 cows, all named after the resorters who came to the market they opened each summer where they would buy fresh produce, eggs, cream, chickens, and home baked cakes. Her mother made cottage cheese every day too. If resorters came in later in the day, they were treated to a glass of homemade wine. Of course, the men were busy with the orchards that made up the biggest share of the property. They raised cherries, pears, and plums. Also out back by the orchards that still exist is one of the biggest chestnut trees on record. Molly remembered, in her youth, that in summer when it got too hot, they would go to the ice house, which was an attached room at the back of the house, and sit on the ice blocks. She said they were cut from ice on Swaney Lake just down the road in the winter.

The barn was the kids' playground, Molly said, and they spent many happy hours there. One game Molly's sons played was called "King of the Mountain." This involved crawling up on the sloping roof of an addition on the back of the barn. Below, on the ground, was the hog pen complete with resident hogs. The boys would scoot down the roof, land in the pen, and run like mad to get to the fence and over it before the "people-eating pigs" got them. They had free access to the loft, playing in the hay and making mazes. Unfortunately, the barn burned in 1947 when the hired man, who slept in the barn, was careless with his smoking and set the barn on fire. It burned so quickly that it could not be saved.

Before electricity came in 1937, the family used rain water and water from the school hand pump just across the road. Ironically, the Betterment Society of Old Mission thought that sidewalks would look nice and be more convenient to the people in Old Mission, so they were installed about the same time that electricity was installed. Though buried or missing in places, that sidewalk is still there and runs about a mile.

After dinner each night, 14 year old Molly and a friend would walk down to the dock at Haserot Beach to meet other friends and, hopefully, some boys. There was a concession stand there where they could hang out. It sold soda, ice cream, and candy and was called "The Wegas" (Indian word for "cherry"). The day the dock burned was very sad for the whole community. Molly said that was in the summer of 1946.

PRATT/ALTENBURG/LEVIN BARN

There was a great deal of excitement when the cows got out, which happened occasionally. Then the family would have to chase them, usually all the way down to the bay. They were pastured out back of the barn in the field and usually would follow the leader home single file each night. She said it was fun to watch their parade.

In her grandmother and mother's time, Indians were still living in the area. The family and neighbors kept a box of clothes by the back door for the Indian families, theirs for the taking. In return they wove beautiful baskets.

Molly laughingly told me that her Uncle Carl, who went to the University of Michigan, wrote his thesis on "Wash Day on the Farm." Molly had so many stories to tell. I wish I could tell them all, but I will end with this one. One year, she said, her folks made "much, much" money on the cherry crop. They had an outstanding year! What did they do to celebrate? They bought two new Ford Thunderbird convertibles; one was canary yellow and the other was robin's egg blue. His and hers cars. What a way to celebrate.

Molly passed away in late summer of 2005. Her son Steve and daughter-in-law Nikki are now occupying the lovely Queen Ann house in Old Mission.

Memories and recollections by Molly Levin. Other information obtained from Steve and Nikki Sobkowski, and gleaned from Michigan Pioneer, *Volume VII, p. 477.*

Pickers in the 1950's

January 18, 1906
Mr. Rushmore and Mrs. Cane drove from Old Mission to the lighthouse last week, something they have not been able to do before some time in January without a great deal of trouble in getting through snow banks, although this time, scarcely finding enough for sleighing. The winter weather has been quite unpredictable from week to week.

COSGROVE/KITCHEN BARN

This 1900 barn is now only a memory, as it was torn down in March 2005 to make room for a road leading back to a new subdivision. It was of timber frame construction with hand hewn beams inside, and it had a metal gambrel roof. It had two levels with a lower level to house the cattle.

Leona Kitchen told me she lived on the farm with her husband Roger from 1935 until 1967. She fondly remembers that she and Roger had a long distance romance from the time she was 9 years old. She was living with her grandmother, and Roger would visit her. He waited for her for eight years! When Leona was 16 and Roger was 22, they married. The newlyweds rented a couple of rooms in the farm house they later bought.

Physical Description
Date Built:
- c.a. 1900

Construction:
- Timber frame bank barn
- Vertical wood siding
- Metal gambrel roof
- Fieldstone foundation

Louis and Carrie Kauer had built the barn and house on the property, but it was owned by Mary Anderson at the time that Roger and Leona bought the 26 acre farm for $1800, borrowing money from Joe Kroupa. Joe is said to have kept his money in quart jars in his barn. She remembers Roger's folks co-signed the loan.

Roger's parents, Mr. and Mrs. Joe Kitchen lived on a farm just North of Roger and Leona's farm. Father and son, working together, farmed about 66 acres of cherries, pears, peaches, and plums. Roger did the tractor work. Leona said they had 26 acres and her in-laws had 40. They transported the cherries into the Cherry Grower's Factory in Traverse City. The families tried to be self-sufficient by canning produce from the garden, along with keeping chickens and hogs. They grew potatoes too. All they bought was flour and sugar. There were some hard times when crops failed, Leona remembers. Her mother-in-law sold eggs and butter to Pete Lardie for his store in Mapleton. Roger also worked for the Kroupas and the Gleasons sorting cherries in season.

Leona and Roger's two children were born at the home of Roger's parents, who assisted with the birth until the doctor could get there to make the delivery. His charge was $25, and a nurse to help afterwards was $5. At the time of the birth of her son Roger and daughter Mary Jo, it was common to stay in bed for two weeks after giving birth.

Leona told me that her husband cut the first trees down to make the road out to the Old Mission Light House. It had been just a narrow two track road for years. He ended up getting a bad case of poison ivy for his effort.

Roger's parents died in 1967 and he inherited their farm. It was hard to get good help to harvest the cherries, so they bought an expensive cherry shaker and used it for five years. But trying to keep government records and managing migrant help was becoming too complicated, so in 1971 they sold the farm to James Cosgrove. The story goes that James Cosgrove stopped to buy cherries at the Kitchen's fruit stand and ended up buying the farm too. Roger loved to fish, so they moved to Rapid City near Elk Lake. They both continued to work at various jobs that were not as taxing as farming. Roger has since passed away, but Leona has a sharp mind and wit, and was a delight to talk to about the "old times."

Recollections provided by Leona Kitchen.

The Gust Seaberg barn in two seasons

THE SEABERG BARN
BY BLOSSOM SEABERG, DILLON, JERUE

As near as I know, it was 1887 when my grandparents left Cadillac, with a team of horses, a cow, and a calf behind the wagon to settle on the Old Mission Peninsula. Both grandparents came from Sweden and met in Cadillac where my grandmother cooked in a lumber camp. She came with her two children, Gust, age 8 and Tillie, age 6. This was 1882. The children's father died when they were very young. There in Cadillac, Jenny met and married Erick Seaberg. They were to have six more children. Gust was my father.

Their first winter was spent in a makeshift log cabin cut into the hillside. My dad told me the first winter was very bad. Their diet was rabbit, fish, and sour milk. The next year was better because they were able to have a garden. The house and barn were built soon after, around 1888, and still stand on Bluff Road, the house across the road from the barn.

As a young man, my dad lived with, and worked for, a man named Sargent. That farm later became the Edmonds farm. In the winter my Dad would take his team of horses and go to the lumber camp and log. With the money he earned, he could then buy the land that became our home, next door to our grandparents' farm. He bought 60 acres with 20 acres along the Bay. The house and barn were built with lumber taken off the acreage as it was 60 acres of woods. This was in 1905 or 1906.

Physical Description
Date Built:
- 1905-06

Construction:
- Timber frame barn double ramped and extensions on both ends
- Vertical wood siding
- Asphalt shingle, gable roof with cupola
- Foundation was probably fieldstone

My mother also came from Sweden, in 1889. She met my father through a cousin. They were married in the small white church in Old Mission in 1907. Mother and Dad had four children. I was the next to the youngest, born in 1912. Everyone worked on the farm and I remember driving the team, milking the cows and tending to the other farm animals. I was a true farm girl!

One day when I was about 11 years old a hobo came to the back door. I was the only one home. He knocked on the door and asked me for something to eat. I made him lunch! I gave him a sandwich and a cup of coffee which he ate on the back steps. When he finished, he tipped his hat, thanked me, and walked away. This was in the 1920's. I wasn't afraid of him or anything, as I would have been today.

My mother had never had her hair cut and wound the long braids around and around her head. As an older man my dad walked very bent over from all the hard work he had done. Mother often said, "It was darn lonesome on the farm." There were only six families living along Bluff Road in those early days…the Edmonds, the Erick Seabergs (grandparents), the Olsens, the Schaeffers, the Fowlers, the Gust Seabergs.

Bluff Road was just a path in the early 20's. There were only four turn outs where cars or horses and buggy could pass. In winter, snow sliding down the bluff was always a hazard to travelers on the trail. My folks, and others, had to drive out on the ice along the bluff to avoid the avalanche of snow that could possibly come thundering down. As kids we skated on the Bay for fun. One year in the 1920's the ice on the Bay must have been a foot and a half thick. When it started breaking up in the spring, it pushed up on the shore to the height of 30 feet, and Bluff Road. disappeared under a mountain of ice. It was August before they could drive along the road again!

I went to Maple Grove school with my brothers and sister. We walked through the fields and up and down hills to get over to Center Rd. where the school was. We did this all winter, unless the snow was too deep; then we stayed home.

> Blossom told me one story that I must retell. When Blossom was a young teen she was helping pick cherries on the Charles Carroll farm on Center Road along with other family children. A real thunder boomer storm blew up suddenly. The kids ran for the barn. As the storm abated, Blossom and the other children ran to the house where they found Mrs. Carroll as white as a ghost. It seems baby Bernie and the cat had been looking out the front screen door together. Baby Bernie crawled away moments before lightning struck the telephone line by the door. The cat had remained by the door, and was killed.

Blossom has outlived two husbands and is 93 years old (2006). She is active, full of spunk, and a delight to be with. She can really tell a good story. The barn on the farm where Blossom was raised was torn down when it was no longer needed. The farmhouse still stands and has been completely renovated and enlarged. It is like a step back in time, 100 years ago, as you drive by this lovely house.

RIDGEWOOD FARM

This phantom barn held the record of being the largest in Grand Traverse County in 1906, when it was built. It was reported to be 100 feet long, 75 feet high and 50 feet wide. I have been told, jokingly, that it was too big to get in one photo. Thus, through the courtesy of Barbara Andrus Bee, I have three photos of the barn to get the whole picture of it. The picture with the children in front of the barn shows just how big that barn was in the 1930's. Also Grand Traverse County encompassed a much larger area at that time.

This large bank barn had both horizontal and vertical siding, a fieldstone foundation and various cupolas. The nailing pattern that is apparent in the picture suggests this was a plank frame barn. It had an interesting side addition coming out from the barn on one side. It was called the work room. There was also an addition on the other side of the barn that was used for buggy washing; yes, buggy washing. A buggy was run up a ramp and washed with water pumped from the Bay.

As time went on, the barn changed. It was re-roofed (in the forties), and the cupolas were removed as a part of the re-roofing. The barn held horses on one end and cattle on the other. There were a few milk cows for the family's personal use, but mostly the barn held feeder cattle. When fruit became the predominant crop grown on the farm, the barn was principally used for storage of machinery and as a machinery repair shop. The roof was lowered in the 1960's as part of the change over from general farming to fruit production and, economically, there was no need for all the extra space in the top level when hay was no longer being grown and stored for the animals.

Physical Description
Date Built:
- 1906

Construction:
- Plank frame barn with side building attached
- Horizontal and vertical wood siding
- Gambrel shingle, later metal roof with cupolas
- High fieldstone foundation

This property has a long history and has had many owners. Parts of this large acreage were sold to different people, only to come together again under one owner. The very first owner registered is Oscar Stevens. In July 1861, he was issued a patent deed, signed by Abraham Lincoln, conveying to him property in Section 23, which is at the end of the Old Mission Peninsula. As often happened in these times, a veteran from some war would be granted a piece of property for service rendered. Often it was not in an area they wanted to live, so they quickly sold it for needed cash. This he did with property being sold several more times.

George Parmalee and his wife, Hulda, bought it by warranty deed (the title warranted it to be free of defects, for example: mortgages) on November 1, 1867 for the amount of $2000. I believe this was about 380 acres. O.H. Ellis was the next owner, as he bought it from Mr. Parmalee in 1883. The acreage was sold several more times, and in 1902, when a former owner defaulted on his mortgage, J.C. Howe and his wife Irene bought the property from the Traverse City State Bank for $12,500. It was Mr. Howe who built the fabulous big barn. Mr. Howe was a patent attorney from Chicago who knew very little about farming. According to David Murray, II, he poured a great deal of money into the farm. At one time the farm had close to 100 heads of cattle. These were registered Shorthorn (Durham) cattle. The farm was a very big operation with a mixture of farming. They raised crops to feed the animals—oats, corn and hay. There was a large building to house hogs on the property. They had milk cows and horses. They had extensive orchards raising cherries, plums, apples and peaches. It was a self-sufficient farm in many ways. But then, it was reported, Mr. Howe did everything in a big way. Bob Tompkins was hired to manage the farming operation. In 1914, the Howes named the farm Ridgewood Farm, Inc. After 1914 the Ridgewood farm was not profitable…partly because Judge Howe was a better attorney than a farmer, it has been said. In the period 1914-25, the Howes entered into a series of loans and mortgages to stay afloat, but in 1925, Mr. Howe lost his property to the banks.

In the 30's the Traverse City Depositors Corp. acquired the property, and from 1933 to 1940 the Andrus family worked the farm for the bank. Barbara Andrus Bee's memories include playing in the hayloft and "walking the planks." It seems there were two boards side by side that ran the length of the barn. The Andrus children would walk from one side to the other of the big red barn. Barbara remembered there were always pigeons in the barn. (David Murray, II, told me that doves were also in the barn. He called it a cote. In fact, they had their own little room with a door to come and go as they pleased.) Barbara remembered her mother and sisters cooking huge meals for the threshers at harvest time. She loved playing in the orchards, making "houses" from the rocks and stones always in abundance in the fields.

In 1939, David Murray, Sr., bought the Ridgewood farm, with help from his father Ben, who was the head of the international

RIDGEWOOD BARN

Middle section of the barn 1930's

North side of the barn

South side of the barn

export department for Montgomery Wards & Company in Chicago. It seems David Murray, Sr., already had a farm on Center Road. David Murray, II, tells the following story about his father and how he became a "farmer."

It seems Grandfather Ben, who was a very good businessman, was concerned because his son David, Sr. could not settle on a profession, having attended several prestigious universities, but never settling on anything in particular. His curriculum usually focused on agriculture. So, his father, Ben Murray said, "I'll buy you a farm for you to settle down on." As David was attending University of Illinois Agriculture School at the time, his father presumed his son would practice what he had learned. David, Sr., moved north in 1916 to a farm on Center Road. Not only did he become a successful farmer, raising fox and growing cherries, but David, Sr. also became a banker in Traverse City. He created the Production Credit Association to help farmers get short-term crop loans that would not be given by the larger lending institutions. It is said that Mr. Murray raised fine silver fox for their fur and exhibited them at the Chicago fur exhibitions. This comes from George McManus, Jr., whose parents worked for Mr. Murray on the fox farm. David, Sr., was also well known on the Peninsula for writing a newsletter during the Second World War. He sent it to all the Peninsula men and women in the service; knowing they would be grateful for some news from home. Sons, David, II, and his brother Ben both served in World War II.

Ridgewood Farm always had a good manager, as none of the Murray family chose to live there full time. In 1943, the Art Wheelock family moved to the farm as manager for the Murray family until 1977. Cattle were phased out, and the concentration was on cherries and other kinds of fruit…plums, apples, and peaches. David Murray, II, said that, for several years in the 1970's, the Murray farms produced more sour cherries than any farm in the world. At the time they were farming 700 acres on the Peninsula. He remembers in 1969 only 25% of the cherries were taken from the trees by shaker and 75% were hand picked. (The family had 4-5 farms at that time.) The Ridgewood farm's acreage ranged from 325 to 514 at various times. They had many, many migrant workers some of which were housed in a large building that was originally built for hogs on the Ridgewood property. It had been built and used in J.C. Howe's time, and had not been used for swine in a long time.

In the early 1930's David, II, said his parents ran a summer camp for "city" kids who came to experience rural life but mainly just to vacation. It helped pay the bills of running the big farm during the difficult times of the Depression. The summer camp was held on the Center Road farm, a large house where two generations of Murrays lived.

David, II, was trained as an English/Social Studies teacher and taught five years in the Traverse City School system and five years at Interlochen Arts Academy. But he confessed he always was more a farmer than a teacher. That was his first love.

When the Murray boys were in their teens, they would go out to the Ridgewood farm with friends and have picnics and parties. They called it Ridgewood Lodge. The boys also lived there at times, from 1975 until 1988. The big barn burned in the 1970's. In 1988 the Ridgwood Farm property was sold to the American Farmland Trust, who eventually sold it to the State of Michigan for a State Park. After AFT acquired the property, the Peninsula Township Fire Department used the house, now in rundown condition, as practice for a controlled burn. No evidence that a magnificent barn once stood on the spot can be found today…making it truly a phantom barn.

Information and pictures provided by David Murray and Barbara Andrus Bee.

Now for an interesting story. Mr. Murray (David, II) was an English teacher in Traverse City.
In his first year of teaching he was working on a unit on insurance…all the different kinds.
A student came up with "twins" insurance.
You see, Mr. Murray and his wife had been blessed with two sets of twins already.
This gave David the idea to insure his wife, Jane, through Lloyds of London against having the third set of twins.
David says they paid $100 for a $1000 insurance policy, a good deal of money in 1959.
The odds were one in 857,000 that Jane Murray would produce a third set of twins.
She did! They collected! They were all boys!

Mr. & Mrs. David Murray, II and their three sets of twins - 1950's

CENTENNIAL BARNS OF OLD MISSION PENINSULA

Cherry sorting—1910-1920
Picture courtesy of Grand Traverse Pioneer and Historical Society

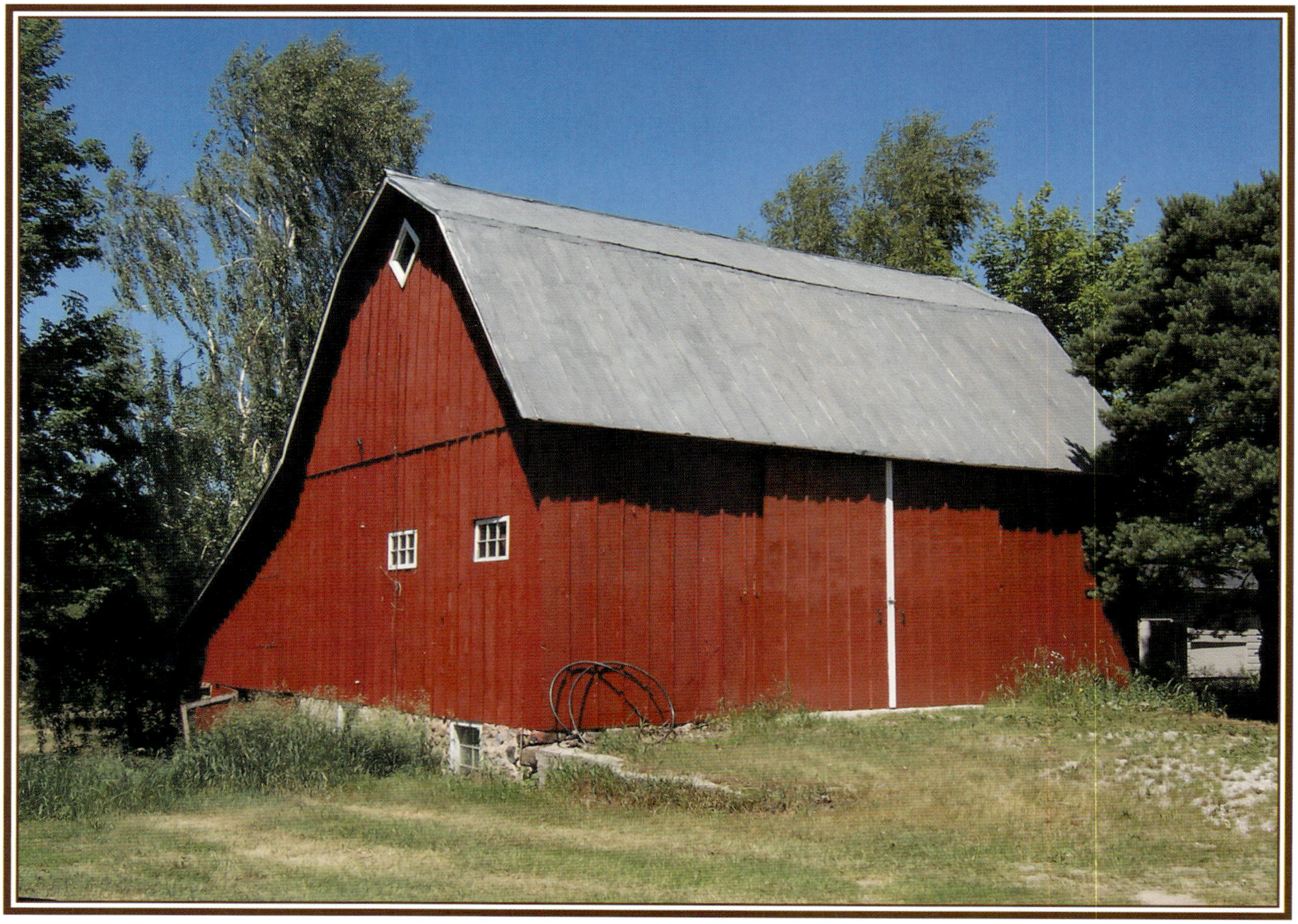

CRAMPTON FARM

This attractive red barn is over 100 years old, built in 1904, and now is a centennial farm. Norm Crampton's grandfather Frank Steavens came by stage coach from St. Johns, Michigan, to the Peninsula in 1895. He went to work for the Walker sisters on Old Mission Road. There he met Mary whom he married. About 1904, the couple settled on the land that is still in the family today, owned now by Norm and Wanda Crampton. They built the house in the dead of winter. Norm said he was told they built a fire in the basement to keep warm while they dug out the dirt using a team of horses and a slush scrapper, or a slip scoop. The barn was built shortly after. Both buildings have stone and cement basements. The team of horses was used to drag the large stones into place.

The gambrel metal roof barn is of timber frame construction with a shed added on the back to give them more room. The barn had a silo at one time where silage for the cows was stored. Also the barn itself had a hay mow in the upper level. The barn was used to house cows, horses, and hogs. There remains an interesting hole in the foundation on the east side where the hogs could come and go at their leisure. Norm does not know how it got there but he assumes his grandfather knocked a hole in the stone wall for their convenience.

Norm Crampton was always a farmer…primarily growing fruit apples and cherries. His Grandfather Crampton was a blacksmith in Old Mission, and his father, Elmer, was an engineer on the Great Lakes during the depression. His mother, Gertrude, and Uncle Charles worked their farm, along with other neighbors during this time to help make ends meet.

Physical Description
Date Built:
- 1904

Construction:
- Timber frame bank barn with extension on back
- Vertical wood siding
- Metal gambrel roof
- Fieldstone foundation

Norm remembers when his father got his first John Deere tractor in 1942. Until then all the plowing was done by horses. "I remember my dad yelling, 'Whoa!' To stop the tractor." Working with tractors took some getting used to. Before cherry shakers came into the picture, the Cramptons had 50-60 Mexican laborers to help with the harvest. Norm and Leland Gore, another Peninsula farmer, were among the first to invest in a cherry shaker in the Old Mission Peninsula area in 1968. It was called a Friday (the maker) and was mounted on the front of a Ford tractor. Tree limbs were shaken one at a time. Later Norm replaced the limb shakers with a trunk shaker called a Shipley. It shook cherries onto a 24 foot tarp that rolled out like a window shade (called a run out). An elevator hoisted the cherries up into a tank. Several years later he got another trunk shaker from the Arnold K. White Company, a West Coast company, that built specialized equipment. Norm had to buy a tractor just for this new cherry shaker.

Norm attended Ogdensburg School and remembers that the school was the first on the Peninsula to have flush toilets! His wife Wanda and he have been active in the community. Norm worked the farm and drove school bus for many years while Wanda was employed at the Peninsula Community Library for 24 years. The couple were also active in Jr. Farm Bureau and the local Farm Bureau Group. People like Norm and Wanda Crampton are comfortable to know. They know the land, the Peninsula, and its people. They know farming and are unassuming in their expectations that almost anything can change and often does. Their three children have all purchased portions of the Crampton Farm and carry on as the fourth generation farmers on this family land. Norm and Wanda are proud of that.

Information provided by Norm and Wanda Crampton.

The hog hole

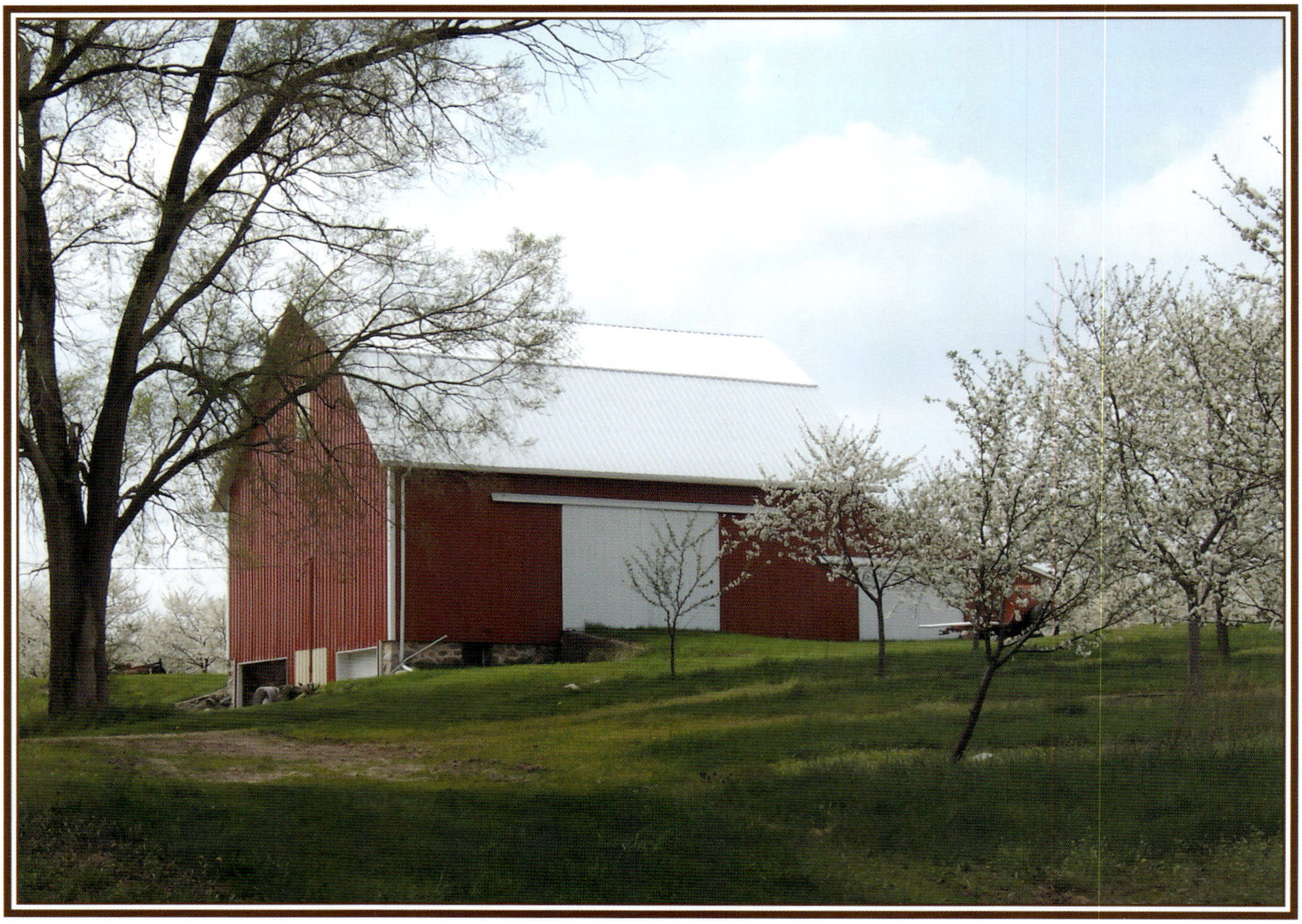

ISLAND VIEW ORCHARDS...THE LYON BARN

The Lyon Barn was built in 1908 and has continued in the family starting with Alfred Lyon, John Lyon, and now Whitney Lyon. It is 36 x 46 feet and stands up near the top of Island View Road. It used to be called Swedish Crossing Road in the early 1900's when many Swedish families came from the old country to take up farming in this area. All together Mary Lyon counted 17 Swedish families that settled in the immediate vicinity. There was once a Swedish Baptist Church just down the road from the Lyon farm.

"Back in 1900 when the land was settled Grandpa Lyon (Alfred) walked to town 6 days a week to the Hannah-Lay saw mill to work off the loan on the farm of 40 acres," Whitney Lyon told me. "It took him two years to pay off the loan given by Mr. Hannah. Then, as he became more prosperous, he loaned money to help the start of Rennie Oil Company in Traverse City!" The farm was a 40 acre potato farm up until the 20's. The potatoes were hauled to Bowers Harbor and two schooners took 5000-6000 bushels to Chicago. The schooners were anchored in the Chicago River and the potatoes were sold right out of the boat. Then another 40 acres were added with fruit trees being planted around 1915. The farm now consists of 100 acres. Whitney remembers that during the Great Depression, cherries from their orchards were bringing 1-2 cents a pound, or 25 cents a lug. Financially, some years were difficult on the farm. Mary remembers there were times when they lived off her daycare money because there was not enough income from the farm.

Physical Description
Date Built:
- 1908

Construction:
- Timber frame bank barn with shed extension
- Vertical metal siding
- Metal gambrel roof
- Fieldstone foundation

Between 1940 and 1971, the barn was used by the migrant families as their living quarters during cherry harvest time. Partitions were put up to create private quarters for at least six families. They used to nail cherry lugs up on the walls to store their food items.

The barn itself held cows and horses until 1950. The beams are all hand hewn and pegged. Men were hired for $1 a day to square the beams. The wood came from trees on the farm. The beams that are the length of the barn are 46 feet long and the beams the width of the barn are 36 feet. Whitney said that 90% of the barns in the immediate area were built by O.J. Benson and his brother-in-law, Ludwig Nelson. (He also said the second Catholic church down on Center Road was built by O.J. Benson.) The barn has the traditional stone foundation and was re-sided with red steel siding in 2000 and reroofed in October 2004. The new roof was installed for $11,880. The grandchildren practiced with their BB guns on the barn windows, so most of the windows were covered up at the time of the residing of the barn. A globe from one of the lightning rods, which was placed there by Alfred Lyon in 1908 when the barn was built, is now a conversation piece on Mary Lyon's buffet.

Whitney, the second oldest of eight children, was raised on the farm. His farm chores were to hoe around trees, harvest fruit and drive the tractor. "At 14 years old we hauled cherries to Traverse City by truck," he said. "There was not the traffic back then and I knew all the people along the route." The family bought their first car in 1915. Prior to that they used a team of horses for all their travel.

The neighborhood was a close knit one and he remembers that when the neighbor's wife died, the other neighbors all pitched in to do his chores when he got sick. The neighbors traded help. The butchering was done by a neighbor. The smoke house for the butchered animals was down the road on the Carroll farm. Eggs and other things were traded for whatever was needed. The farms, in this way, were self sufficient.

The Lyons grow eight different types of cherries and eight different types of apples, besides peaches and plums. And they have raised all varieties of children! Besides having had ten children of their own, Mary still does daycare in the home and helps out with grandchildren too!

Five generations of Lyons have worked on the farm. Whitney and Mary's son Frank does a great deal of the farming now though Whitney continues to farm. It seems every time I came to the farm to talk to Whitney, was working...picking apples, or fixing something. Mary calls Frank their right hand man. Grandsons come to help on the farm too. Grandson, Jack, a teenager has a "farm work ethic", and is a very hard worker. Whitney jokes that leaving a farm to your kids might not be considered a gift because of the uncertainty of farming.

Information provided by Mary and Whitney Lyon.

Picture courtesy of *Barns and Blessings*—Robin Grothe photographer

WALTER and MARY JOHNSON BARN

Driving south on Center Road, shortly before you come to Mapleton, you round a curve in the road and there, right next to the road, is a lovely saltbox style red barn with white trim. It is a picturesque scene with fields of cherries behind it and the Bay in the distance. Another curve is right ahead so you cannot stop and take a picture. Let me tell you about the barn and its family.

The Johnson barn was most likely built prior to 1880 by Robert Edgecomb. Walter Johnson's mother was married to Mr. Frank Edgecomb, son of Robert Edgecomb. After Frank passed away, Stella Smith Edgecomb, his widow, ran the farm. The farm was a land grant given to an owner before it was purchased by the Edgecombs. Mr. Edgecomb's trademark was that he built with fieldstone, and he laid the fieldstone foundation of the barn. Originally, the barn had a silo on the corral side, which toppled partly, probably in a storm. The rest was taken down. Later the silo was moved to the inside of the barn. The cupola blew off in a storm, also.

Stella Edgecomb married Lester Johnson in 1919. Her first husband had died of influenza in World War I. Their oldest son, Walter, who had a Masters degree in electrical engineering from Northwestern University, came to settle on the farm and take up farming when his father retired because, as the oldest son, he inherited the job.

Physical Description
Date Built:
- 1880

Construction:
- Timber frame bank barn with side addition
- Vertical wood siding
- Metal gabel roof
- Fieldstone foundation

Walter was a man of many talents and interests, and was not only an engineer, but was a lieutenant in the Navy in World War II and in the Coast Guard reserves. He also became an avid collector of Old Mission Peninsula history and he was known by all on the Old Mission Peninsula. While at Northwestern, Walter met Mary. She was a telephone operator on campus and, although originally from Virginia, she was living with her family in Evanston while her father worked on the Great Lakes for the Army Corps of Engineers. They were married in 1946. Mary was in her last year of nursing school back in Virginia and Walter was still in the Navy. When they moved to Old Mission Peninsula, Mary had a lot of adjusting to do. She found it hard to be so far away from family. Carol, her daughter, said, "Can you imagine a cultured southern belle coming to the north woods country and becoming a farm wife?"* Talk about adjusting! They lived in the house across the road from the barn between 1946-1960. There was no indoor toilet in the house when Mary came to live on the Peninsula. When she was several months pregnant, her father-in-law put a toilet in a small dark room under the stairs. Being very frugal, he neglected to consider adding a light! They moved to their house in Old Mission in 1960. The house was then used by their Mexican workers and accidentally burned down in the 60's.

This barn resembles a saltbox-style barn because of its dramatic slant to the roof on the one side. However, it is a gable roof barn with a three sided shed, put on sometime later. Under this addition there is a 1000 gallon cistern where rain water was collected to water the dairy cattle which were kept here until the 1950's. Carol and Dean Johnson also had horses while growing up on the farm. The barn stood empty from 1960-1970. Sturdy as the barn appears, Dean Johnson remembers the barn rocking in the wind in a storm.

Mary remembers that Walter's mother was a bee keeper and besides tending her hives of bees, she also had the job of getting the bees out of Bowers Harbor Inn's chimney one year.

Grandma Johnson

The Johnson farm consists of 80 acres, divided in half by Center Road. Mary remembers trying to bring the cows across Center Road each night. They pastured in the meadow behind the house on the west side of the road. She said you never knew when a car would come whipping around the curve when the cows were mid-road. The family raised cherries and utilized approximately 100 migrant pickers before cherry shakers came into being. They stayed in tents early on, and then had a long building that consisted of several individual rooms, with a hose for water, and a wringer washing machine. The outhouses were up the hill. At the end of the season Mary remembers the migrants, mainly from Texas, would stage a big celebration with lots of good Mexican food. Carol said there would be a huge bonfire that went with the celebration, the highlight for the farm family and the migrants alike.

Daughter Carol remembers that after the hay was baled and put in the barn, her brother Dean and she would make tunnels with bales of hay. They would twist and turn and sometimes they made them 2-3 stories high. She also remembers having a rope swing that was long enough that it would let them swing right out the barn door. Mary and Carol said with a bit of a smile on their faces, that it was a "known" fact that you must keep hay bales stacked in your barn…or the barn would fall down!

This farm is in the Michigan Agricultural Environmental Assurance Program. A few head of cattle are currently calling the barn "home."
* See *Life of a Farm Wife* written by Mary Johnson, her recollections of what life was like on the farm.

Information provided by Dean Johnson, Mary Johnson, and Carol Lewis.

A story on the following page is one that Carol wrote as an adult, remembering what life on a farm was like—to work and to play and to imagine…

Center Road unpaved - 1930's

The present Center Road has been known by various names according to its use in a particular time period. Originally named the Mail Road, it later became the Stage Road, Telephone Road, Old Trail and Queen's Highway (in honor of the Cherry Queen). Picks, shovels and horse drawn scrappers and snow plows were the basic road equipment until the 1930's.

QUEEN OF THE CIVILIZED WORLD
BY CAROL LEWIS
(Copyright by Carolyn Johnson Lewis)

This morning my father drove the Case tractor, raking the cut hay into rows. My mother stood the baler, a rectangular tunnel on wheels, and my brother and our hired man stabbed ice hooks into bales she left behind, and in one strong swell swoop they swung the bales' weight onto the bed on the '59 Chevy flatbed truck where, in the cab, I rose from the driver's seat to reach the starter on the floor. In the southeast corner, in the sunshine that turned the meadow yellow, two Guernsey cows stood ankle deep in the pond.

Three stories in the air now, under the creak of rafters, in the 107-year-old barn that was once our grandmother's and is now our parents', while a young screech owl sleeps on a high beam near us, my brother and I heft, with effort, one bale of hay to rest sideways on top of two others, creating a tunnel below with room for only one of us to wriggle into. I am eleven and quick into it, crawling down under to lie beneath.

Over my head now, my brother jumps on the bale above me, complaining mightily that most if not all of the labor to build this hay castle was his (pound, pound) and therefore he deserved to crawl in first (pound, pound). Mr. Irritable, Mr. King of all Russia, Mr. I-Want-What-I-Want: hay tunneling takes two.

And so, when I am done considering where I will place a TV in the three inches of space I allot for it, and where the big armchair, in the four inches of my left-hand corner, and where my right toe is poked by hay might go my new 45 rpm record player I got for my tenth birthday, then and only then, shall I decide whether I'll crawl out and let my brother come in, or whether I might just take my afternoon nap.

And while I am under the hay listening to his ongoing harangue, a monologue no less in words than any president's first address, it is rest that takes over, sleep sifting into weary muscles, until now, in the dusky shadows of late afternoon, I close my eyes and dream. I am a soldier-girl in a war against cowardice, Queen of the Civilized World. With honor, I fight the ignorant, the uninformed, those who slaughter innocents. My horse, clad in mail, responds to my shifting weight, stepping, turning sideways, swinging round at the touch of my knee as, courageously, we banish those who threaten the future of our young.

It is dusk when I wake. The dinner bell is clanging for the family and farmhands to wash up at the outside tap. I take my time: flicking off my imaginary TV in the corner, turning the knob on the record player so the disk—that isn't there—slows and stops. I back out from under my hay bale, yawning and stretching with the soundest of sleeps, and sink back onto my heels. Across from me, on the high barn beam, in the dusk, the yellow headlights of the screech owl's eyes ponder me. As if in some complicity, he winks once, and lifting his massive wings flies up, up through the dusk near the rafters, then turning, he wheels, swooping out through the barn doors flung wide, nearly brushing the head of my brother who, standing in the doorway, hands on hips, brushes off the wingbeat with a furious arm's arc, his foot impatiently tap-tapping, tap-tapping: Mr. Irritability, Mr. Miffed, Mr. King of all Russia.

"Aren't you coming?! She's called you three times!"

And though my body will brush past the enemy and walk across to the house for dinner, my horse and I, and our trusty owl, have just broken through the front lines…

Bud Kroupa with the axe that built these buildings

One of the few remaining windmills on the peninsula

BUD and NAOMI KROUPA FARM

Following the old Indian Trail up the Old Mission Peninsula, Bud's great grandparents, Leopold and Petronella, came to Bowers Harbor in 1852. They found what they were looking for…beautiful forested country with much potential for farming. It had been a long journey from their village in Czechoslovakia to this destination. With them came their four sons, Charles, Ferdinand, Loudvik, and John. Ten of their children had died of scarlet fever in the Old Country. Life was hard for them in the wilderness, as Leopold had owned and operated a large hat factory in the Old Country, and his wife was from an aristocratic family. They were not farmers. It is said they came to this country to keep their four sons from the European army draft. However, Charles was called up, passed military examinations, and was inducted into the army during the Civil War.

Physical Description
Date Built:
- 1859

Construction:
- Horizontal on grade log barn
- Vertical siding at intervals
- Metal gabel roof
- Fieldstone foundation

I dare say all of the Kroupas on the Peninsula today are descendants of these Kroupas. Bud Kroupa, whose homestead has been in the family for 146 years, can claim Charles as his grandfather. His father, Albert, was the ninth of ten children born to Charles and Mary. He was called Harbor Bert as there was another Bert Kroupa on the Peninsula and he lived near Bowers Harbor. The other Bert was a son of John Kroupa, and he was called Frisky Bert. But that is another story for another barn! The Kroupa family got their land through the Federal Land Office. Abraham Lincoln signed the land grant as he did for so many others on the Peninsula.

When the first Kroupa came to the land, it was totally covered with forest. Bud believes that they built the log house first, in 1858. As they cleared the land, they planted cherries, apples, and, in the early days, corn and hay. They had cattle and hogs. The barn, a log barn built in 1859, is truly unique. It is one of the oldest barns on the Peninsula, and it is the only log barn. It is constructed of hand square logs with dove-tail joints. The horizontally placed logs came off the land, and Bud showed me the broad ax used to shape the logs. "Witness" marks from the ax are still present on these 16-18 inch thick square logs.

Dove tail joints on log house

A part of the original house

BUD KROUPA FARM

Right outside the back of the present house, which was built in 1913, is a large root cellar, where produce stays fresh all winter. In the backyard is a chicken coop, a granary, a smoke house, and a complete windmill. Each end of the barn has a three wall shed that housed the hogs on one side and the milk house on the other. One wing of the original log house is still standing in front of the barn. It was used as a blacksmith shop after the frame house was built. All of the buildings are in good condition and have recently been painted a nice bright red. It comes to mind that this barn and buildings look like a museum of a working farm 100 years ago!

One interesting tidbit that Bud told me was about his grandmother, Mary Kroupa. When she had lived in the U.S. for five years, she went to Chicago to get her citizenship papers in 1865. She saw Abraham Lincoln's body on the train carrying him back to Springfield, Illinois, for burial after his assassination.

Bud told me that many Indian arrowheads have been found on or near the Kroupa property as they would come from Peshawbestown on the Leelanaw Peninsula to hold powwows in the early 1900's. Sometime before 1871, a small pox epidemic raged through the Indian villages. His father told Bud that when a family was stricken with small pox, and everyone in the family died, the Indians would burn the dwelling and all that was in it in an effort to stop the epidemic. He said his father remembers looking across the Bay and seeing two or three fires at a time. It nearly wiped out the Indian population on the Leelanau Peninsula.

The Kroupas' picture perfect homestead is on Neah-ta-wanta Trail, an interesting, obviously Indian, name for a road. The road got that name as a result of all the vacationers and resorters that traveled from other states by boat, train, carriage, however they could get to this wonderful natural area for a respite in the summer. (Many of the vacationers were from Cincinnati.) They eventually formed an association as they began to buy and build summer homes, and needed a name for the road. It was called Nee ah we da ta, an Ottawa word for "I'm going to visit for awhile." It was eventually changed to Neah-ta-wanta.

There are 53.4 acres on the farm site today where Bud lives with his wife Naomi. In 2005 they celebrated 63 years of marriage.

Information by Bud and Naomi Kroupa and from The Descendents of Simon Kroupa by Robert Ellis.

Kroupa root cellar

1904-05
On Friday evening while Albert Kroupa and several family members were out pleasure boating at Bowers Harbor, they saw a monster fish near Johnson's dock. A spear was procured and the fish proved a good mark, but in the efforts to land him, the boat almost upset and the attempt was abandoned.

KEITH and JEAN WARREN BARN

This homestead has been in the Warren family since 1867. Its first owners were John and Rebecca Warren, then Frank and Esther Warren. Next it passed to Keith and Jean Warren. Keith passed away in 2005, and the farm will continue in the family with son Gary. This makes it a fourth generation farm. The 26 x 62 foot barn was built in 1885. It was built, with neighbors' help, from lumber taken off the farmland. It is a timber frame barn built into the hillside and has a fieldstone foundation. The barn was moved back from the 2 track road when the road was to be widened in 1930. Two teams of horses were used to roll the barn on logs. The house was moved too. Until the 1950's, the lower level of the barn housed cattle, horses and hogs. Now it is used to store hay, corn, etc. The barn was added on to at the back at one time. There is a nice gable roof corn crib standing in front of the barn.

The farm now is a cherry farm. The acreage started out as 60 acres and has grown to 85 acres. The farm was once self-sufficient except for flour, coffee, and sugar.

In horse and buggy times, the Warren homestead was used as a half-way house. Wagons stopped to water their horses before they went up Kelly Hill. I am told that Kelly Hill Road used to go around the other side of the hill. The Warrens had a small store that sold a few items such as fruit, vegetables, and tobacco to those who stopped. Now they sell apples in the fall from a stand in the front yard.

Physical Description
Date Built:
- 1885

Construction:
- Timber frame bank barn with ramp
- Vertical wood siding
- Metal gable roof
- Fieldstone foundation

Keith was an outstanding farmer on the Peninsula and was an active member in many of Michigan's fruit industry organizations. He received numerous awards all dealing with his outstanding service to the fruit industry. Keith had two brothers. Elmer farmed all his life as Keith did. The other brother, Ebb, went to work in town and never took to farming. When he first got a job in town, he proudly brought home a gift for his mother—a radio. Keith thought it probably was one of the first radios on the Peninsula. Ebb's mother thanked him but said she would rather have had a sack of flour. Ebb went on to become a celebrity. "He is the famous Ebb Warren, DNR writer and photographer known state-wide as one of the early pioneers in natural resource conservation in Michigan" quoted from writing by George McManus, Jr.

Keith Warren remembered a happy childhood with his brothers. He said everyone shared the work. He had a real love of the land.

Oral information by Keith Warren.

LARDIE BARN

The Lardie family came from France to Montreal and then to Old Mission Peninsula after seeing government brochures that showed what land was available for homesteading in 1851. Oakley Lardie told his interviewers, Walter Johnson and Julianne Meyer, they came over to East Bay by sailing schooner and were just dumped out in a small boat. There was not much land available for homesteading but they found the land that had been assigned to them, and after cutting down a few trees, they built a log cabin and started housekeeping. They had 80 acres. My grandfather, Oliver Lardie, had a trade as a cooper and made barrels which were quite in demand for shipping apples. In this way he was able to supplement his income and acquire a cow. Oakley said the family was one of the first to have a deed to their land. He said many original settlers did not have deeds as the settlers just came and started farming land not yet opened for public sale by the government.

It seems this was not the original Lardie homestead. Oakley's Grandpa Oliver's original farm was just north of Mapleton. He wanted a farm that was down Center Road a couple miles to the south. No deal could be struck with Mr. Eugene Umlor, the owner of the desired property. But Oliver and Mr. Umlor each owned one ox, so finally Oliver offered his ox in trade for the farm and a deal was made. Now everyone was happy. Oliver had his new farm two miles down the road, and Mr. Umlor had a pair of oxen. This story has been widely told through the years, and Oakley loved to tell it. An interesting note: Oliver and Elizabeth Lardie donated a small area of land just north of the present farm for the construction of the first Catholic Chapel (Corner of Blue Water and Center Road). St. Joseph Chapel was built in 1880. It is said that many pioneer families helped build this chapel. Several other churches were built on this same spot over the years.

Physical Description
Date Built:
- 1904

Construction:
- Timber frame bank barn
- Vertical wood siding
- Asphalt shingled gambrel roof
- Fieldstone foundation

The main barn on the property is a 36 x 48 foot gambrel roof two level bank barn and is built into a hill, making a lower level for the livestock It was built in 1904 and it has been used by the Lardie family since then. There are two other out buildings with dormers that are equally well kept as the barn. The barn has lightning rods on its roof top.

The family grew corn, potatoes, beans and other crops. The Lardies had 6 cows, 4 horses, some pigs, and chickens. Cherries and apples became the cash crop in the 1930's. The Great Depression was bad for many on the Peninsula. Oakley remembered they had food, as theirs was a self-sufficient farm, but many did not. His father, Charles, was a relief agent for the Peninsula during the Depression and gave out sugar, flour, lard, butter, and other commodities to those less fortunate. Oakley worked for the WPA and his job was to look for barberry bushes…which caused wheat rust on crops. It grew 7-8 feet tall and they had to put salt on it to kill it.

In the late 20's tragedy struck the family. The family contracted TB from one of their milk cows. Two daughters, Nancy and Catherine, and one son, Leslie, died. Oakley did not get TB, because he did not like milk. Another brother, Stanley, died of TB much later in the hospital in Grand Rapids. While Stanley was in the hospital, Charles slipped and fell in the draft horse's stall and was hurt badly by the horse. He managed to crawl up the hill to the house for help. Stanley died one week of TB and father, Charles died the next week from his injuries. At the time of the tragedy, Oakley, Jill, and their four children were living at Bowers Harbor beside the Harbor Grocery which they owned. Oakley was in the Sea Bees in Guam during World War II and received a hardship discharge from service when he was needed at home. They moved up to their current house after his father passed away, and Oakley took over the family farm.

An interesting note: The land where the men played baseball, now called Bowers Harbor Park, was about to be sold and cleared of all trees. The township voters turned down buying it every time it came up for a vote. With the threat of losing their playing field looming large, Charles Lardie, Oakley's father, borrowed a thousand dollars for ten acres and bought the field to keep it in the community. In 1939 Peninsula Township acquired money from a government grant and other means and bought the park from Charles Lardie. (This information is from *Memories Hidden, Memories Found* in an interview with Oakley Lardie in 1994.)

Oakley remembered walking to Bowers Harbor and diving off boats docked there. He said that was a lot of fun. Social gatherings were card games, pot luck dinners, box socials, and dances at the community hall. He said you knew all your neighbors and it was a close knit community. You never needed to go to town (Traverse City) to find something to do.

Nancy Uithol, one of Oakley's daughters, remembers being a tomboy along with her sister Judy and brothers Ken and Chuck. They would climb to a window near the top of the barn and jump into the hay mow. She remembers having picnics with Grandma Kittie Lardie, probably under one of the old, old maple trees sprinkled around the farm yard. Grandma would churn butter and make bread. Nancy remembers going out under a tree and eating their treat…a real picnic. Grandma Lardie was very progressive Nancy said. She had one of the first GE refrigerators with a round condenser on top. She had a Hoosier cabinet with the built-in flour bin and little drawers for spices. These are all pleasant memories for Nancy.

This is a centennial farm having been in the same family over 100 years. Oakley passed away in the summer of 2005, at the age of 95. His second wife, Mary, still lives on the farm.

Information provided by Oakley Lardie, Nancy Uithol, and from the book Memories Hidden, Memories Found.

The Centennial Farm Program

In 1948 the State of Michigan initiated the Centennial Farm Program which honors the ownership of farms that have remained in the hands of and have been operated by members of the same family for 100 years or more. The qualifications are:
- It must be a working farm.
- It must have a minimum of ten acres.
- Ownership must have remained in the same family for more than 100 consecutive years.
- The relationship of the present owner to the owner of 100 years ago must be clearly stated.

The original timber inside the barn

RILEY/WILSON BARN

The original Wilson farm was on the road that bears the family name, Peter Wilson having purchased 40 acres in 1870. Mr. and Mrs. Wilson had six sons when they came to the Peninsula. One son, William, or W.T., got the farm. William had two sons, Willard and James, and in time, the farm went to Willard. Mrs. Willard Wilson (Mildred) was respectfully known as Mrs. W.F., according to George McManus Jr. "She was definitely the dowager type and was the leader in preservation of the character of the Peninsula. Single handedly, with no laws, rules, ordinances, or bureaucrats involved, she kept signage of any type off the Peninsula. If anyone even thought of putting up a sign, they got a visit from Mrs. W.F. After a little tete-a-tete, everyone backed down." Their daughter Mary married Joe Riley, who owns the farm today. It has been a centennial farm since 1974, in the family and continuously farmed since the 1870's.

This old barn is encased in a green metal siding to preserve it and make it more maintenance free. However, inside it is truly a grand old barn dating before 1874 with hewn square timber beams. Helen Vogel said her great-grandfather Peter was a "square timber" man. This barn had a stable with a cement floor on one end. The threshing floor was of wood with a loft for the hay. The Wilsons were a self-sufficient farm primarily, as they always had livestock, two cows, and horses. There was a granary, with a root cellar under it. In the 40's and 50's there was a pasture out back of the barn for the livestock.

Physical Description
Date Built:
- *c.a. 1870*

Construction:
- *Timber frame with three sided addition*
- *Vertical metal siding covering old wooden barn*
- *Metal gable roof*
- *Concrete foundation*

The Wilsons started growing fruit early and, because of the farm's unique location, rarely had a year without a crop of cherries. Joe Riley explained that the farm is situated between the two Bays where the Peninsula narrows, making the growing conditions almost ideal. The closeness of the Bays insulates the land from the harsh weather and the frosts that plague other Peninsula cherry growers. Joe believes there were only two years in the last one hundred years when there were no cherries at all on the farm. But he points out that they always had to work hard to maintain good orchards and there were hardship years, especially in the 1950's.

Joe said he was not always a farmer. He had an executive job in the steel industry in Detroit in his younger years. Mary, his wife, being a farm girl in her youth, yearned to return to her roots, even though she and Joe came up in the harvesting months and helped out. In the spring of 1970 they left the city life and moved permanently back to the farm. Joe said Mary's father, Willard, and Uncle Jim Wilson were always working to teach him their farming skills during those summer visits so he did not come totally unprepared to be a fruit grower.

The farm at one time had 125 migrant workers, 98% being Mexican. The Wilsons believed in treating them fairly and had a camp with cabins, many of which are still in place today. However the cherry growing and harvesting business was changing and soon the mechanical harvesters made it unnecessary for the large number of cherry pickers. Joe also noted that state regulations and environmental issues were hard to keep up with. When Uncle Jim decided that he did not want to adapt to the new techniques, he retired from farming his land, and Joe now farms his acres of orchards. These 30 acres now belong to Helen Vogel, Jim Wilson's daughter.

The Rileys, Joe and second wife, Ann, have other fruit on the farm. They have 30 honey crisp apple trees, some plums, nectarines, and pear trees which he said are "house fruit" for the family's use. There used to be a 2 acre raspberry patch out behind the barn also, and benches along the edge of the patch for sorting berries.

Information provided by Joe Riley, Helen Vogel, and George McManus Jr.

According to Gil Uithol, one of the township assessors, there are about 91 general purpose barns (meaning old buildings) on the Peninsula. He also said there are 200 farm parcels representing about 9000 acres of farmland. This would represent about 50 long-time farmers working this acreage on the Peninsula.

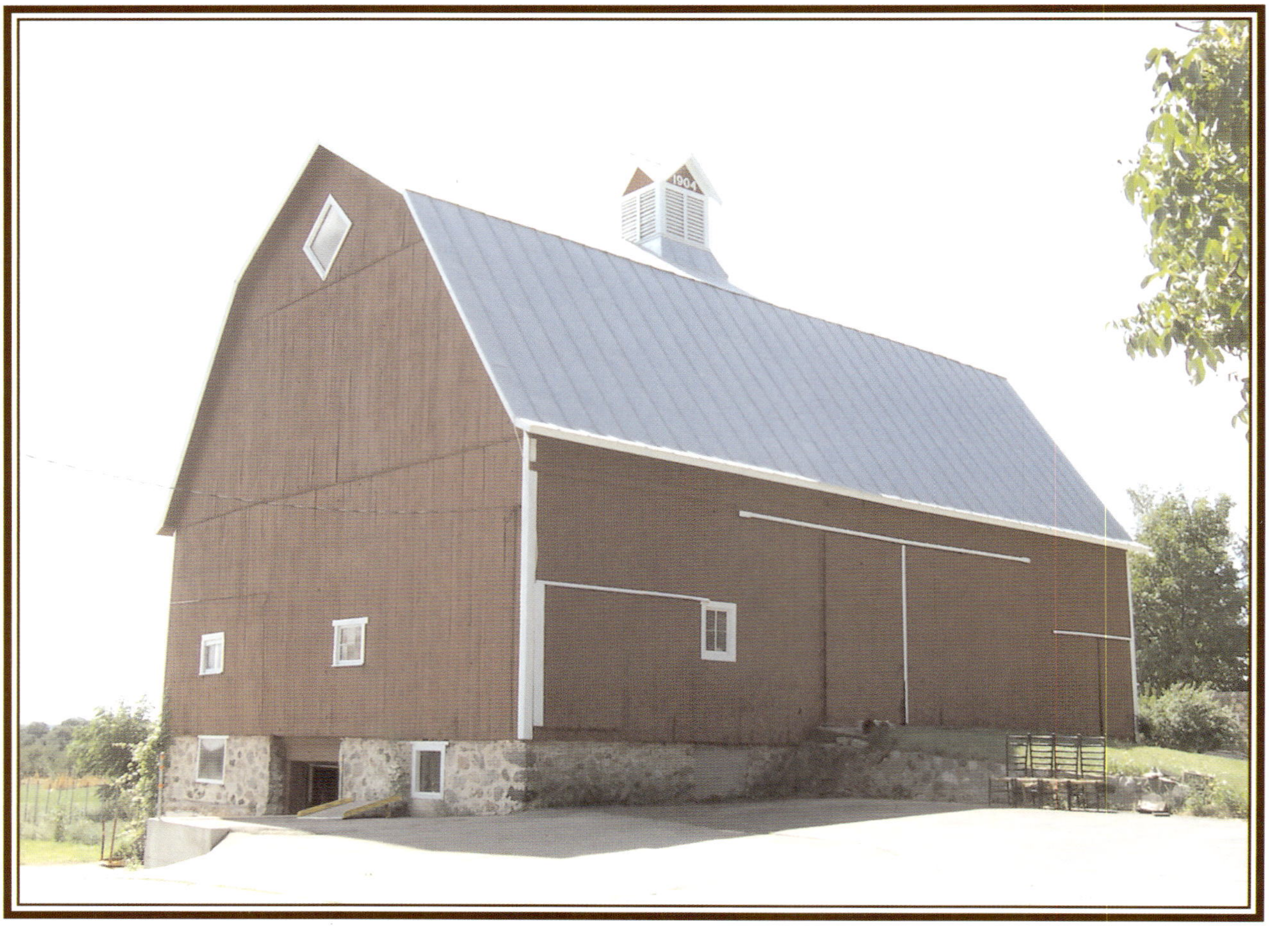
1904

GRAY/SPRINGER BARN

This farm has quite a history. It has been a centennial farm since 1968. The barn was built in 1904. It is 44 x 60 feet. It is a sawed timber bank barn and is truly a magnificent sight. It stands on high ground and can be seen, with its metal roof gleaming, from the top of Carpenter Hill as you start driving out the Peninsula. As you look to the left and the right you can see an azure blue bay on either side. Boaters use it as a reference point when on the water.

It all started in 1868 when Albert P. Gray and his wife Elizabeth moved to the Peninsula. He was an educator who promoted the improvement of schools all his life. He taught for 10 years in Grand Traverse County and held various positions dedicated to raising the standards of schools. Alan Gray, a great grandson, told me that Albert P. donated land for a school which was located where the community of Archie was. It is now called Archie Park. The school burned after a short while and thus little is known about it. The Senior A.P. Gray did teach school at Stoney Beach from 1879-82. Alan Gray said his great-grandfather was a real "Champion for Education" and that many of his ancestors are well educated because of this.

Physical Description
Date Built:
- 1904

Construction:
- Timber frame bank barn with ramp
- Vertical wood siding
- Ribbed steel roof with cupola (date 1904 on it)
- Fieldstone foundation

First and foremost, Albert P. Gray was a farmer. I quote from Sprague's Biography now. "In October 1868, he settled upon the farm purchasing 120 acres of land, the greater part of which was covered with a dense growth of timber. In the midst of the green woods, however, he established his home and began the arduous task of developing his farm. Acre by acre the land was cleared and in the course of time was transformed into productive fields. He also extended the boundaries of his place until it comprised a quarter section (160 acres). 100 acres is now improved. He is a leading horticulturist of his county and his orchards comprise between 40 and 50 acres." There is more about his long and productive life in the biography on him in Sprague's book. (p. 774-5) Barb, Albert's great-granddaughter, said he also raised American chestnut, hazelnut, black walnut, and English walnut trees.

Barb's grandfather, Edwin P. Gray, like his father raised fruit, corn, nuts, and other crops on the farm. He passed away in 1948. Lewis, Barb's father, likewise was a lifetime farmer with 120 acres mainly in cherries and apples. Barb and husband Jerry, a retired educator, now farm the land. So, in 137 years, it seems this family history has come full circle.

In Barb's younger days, the barn was full of loose hay and it was fun for her and her three siblings to play there. But, she remembers being seriously admonished by their father because of the danger of accidents romping in the hay. She said that all the kids had horses when they were growing up. They were active in 4-H and as kids were an integral part of the farm. The kids milked cows, drove tractor, and always had chores to do.

Barb told me that her father (Lewis) raised cattle, horses, sheep, and pigs and housed them in the lower level. Beef cattle were wintered there too. Barb remembers that one year a pack of dogs got loose and ravaged the Gray's flock of sheep, killing them all. While the sheep shed is still standing, the family never raised sheep again.

The barn is now used mainly for storing orchard equipment. The family also had a gift box business at one time, shipping boxes of apples, English walnuts, and maple syrup all over the United States. All of the products were from the farm. This farm was the first on the Peninsula to build a cold storage unit in a building on the farm.

Barb said her heart and soul are in the farm, and it shows. The barn is newly painted and is very well maintained. Recently 101 acres of the Springer farm property were put into the PDR (Purchase of Developmental Rights) program.

Information provided by Barb Gray Springer.

The wellhouse

Old Mission Peninsula has many maple trees tapped for their sap at maple sugaring time. Along Center Road, the Old Mission Historical Society has planted hundreds of trees, mainly maple, but other kinds too, to bring back the character and scenic beauty once found along side of the road.

LESTER BUCHAN'S SUGAR BUSH IN THE 1980'S

Maple sugar making:
In the early spring, as soon as the freeze was broken and the sap began to run, the trees were tapped. A brace and bit drilled a hole 4-5 inches into the tree and sap would begin to run out. In earlier years, a tap (hand carved plugs with holes forced through them and notches to hold the pails) was inserted into the drill hole. The pails would fill overnight…sometimes twice a day. Today, a tap is put in the tree and hooked up to a long plastic hose that runs to each tree and on down the row of trees to a barrel. The sap is boiled down at a sugar house, the moisture evaporating until only the sweet syrup is left. It is prized today, as much as it was 100 years ago, for the marvelous taste used on pancakes, ice cream, and all sorts of other things.

Pictures courtesy of Norm and Karen Buchan

Dinah tending the evaporator

RENOVATED OR RESTORED BARNS OF OLD MISSION PENINSULA

Anam Cara Farm

Restored barn with black angus cattle

Before restoration

ANAM CARA FARM

This colorful old barn sits high and proud out in a large field, guarding the fruit trees and vineyards nearby. It is a pleasant sight to see this nice red barn and lovely home of Tim Quinn and Michelle Keith as you drive down Center Road. The couple bought the 65 acres in 2000 and have spent four years restoring the barn. It is on record as being built in 1915, but it appears much older when you see the inside. The beams of this 30 x 44 foot barn look hand hewn. Part of the floor in this two-level barn is dirt and part has a cement floor. There is a loft on the west end of the barn and a unique hay chute added on the east end of the main level allowing hay to be pitched down to the cattle below. The siding is rough sawn lumber. The roof boards on the inside show a real variety of widths in the lumber used. On the outside the barn has a metal roof. At the top of each peak of this gable end barn is a spirit hole in the shape of a star. I have been told that it protects the barn from evil spirits, allowing them to escape. It also provides a bit of ventilation for the top of the barn.

The lower level housed horses at one time and the stalls remain. On the back of the barn is an added concrete structure where the owners keep three black angus cattle. At one time migrants were housed in the main level of the barn. Today there are 10 acres in cherries and 5 acres in grapes. The acreage is in the Purchase of Developmental Rights Program. This farm was known as the Spruit farm for many years, and I found their name on the 1908 and the 1930's plat maps. Cub Spruit was Central High School's football coach from 1921-27, and a farmer. He had 30 acres of orchards.

Tim and Michelle have named their place Anam Cara which means "Soul Friend" in Gaelic.

Most information provided by Tim Quinn and Michelle Keith.

Physical Description
Date Built:
- 1915

Date Restored:
- 2000-04

Construction:
- Timber frame
- Vertical wood siding
- Metal gable roof with a star-shaped spirit hole at each peak
- Concrete and stone foundation

Front view of barn

Before restoration

FREDERICK BARN

This lovely old bank barn is 32 x 38 feet and was built in 1910 by O.J. Benson. The barn has a beautiful granite stone foundation and a metal gambrel roof. Rick and Peg Frederick purchased the barn in 1989. The barn was needing repair and paint when the Fredericks bought it. They have taken great care with this barn, and it is in immaculate condition inside and out. The barn, which once housed cattle and horses, now has horses living there again. The Fredericks' daughter Meghan was 11 years old when she received her horse, Mandy, as a Christmas present. Mandy was boarded until there was a home for her in the barn they now own. Meghan is now 22 years old and the horse is still in the family, along with Willy, her pasture buddy. What a lovely sight to look at the white fence meandering up the hill with two well cared for horses looking out at you. It is indeed a very peaceful setting. The Fredericks have 6 acres, with a lovely house, up on the hill behind the barn, built in 1990.

Andy Carroll, who was one of the seven sons and three daughters born to Edward and Jane Holman Carroll, was the first owner recorded on this property. All of the sons had cherry farms between Carroll Road and Blue Water Road. Each of the boys made their grub stake by logging in the winter. Tom Hoffman remembers cutting logs with Andy in the winter. Andy got the 30 acres at the corner of Island View and Center Road and the 1908 plat map shows the property in his name. Andy and his wife, had two children—Albert and Helen.

Physical Description
Date Built:
- 1910

Date Restored:
- 1989 to present

Construction:
- Timber frame bank barn
- Vertical wood siding
- Metal ribbed gambrel roof
- Cut granite stone foundation

Smokehouse

Albert ran the farm after his father passed away, and Helen Carroll, who never married, lived in the house on the property. It was a working farm for the Carrolls, growing potatoes first and gradually moving to fruit, apples and cherries, in the early 1920's. The barn stands at the back of a nice apple orchard.

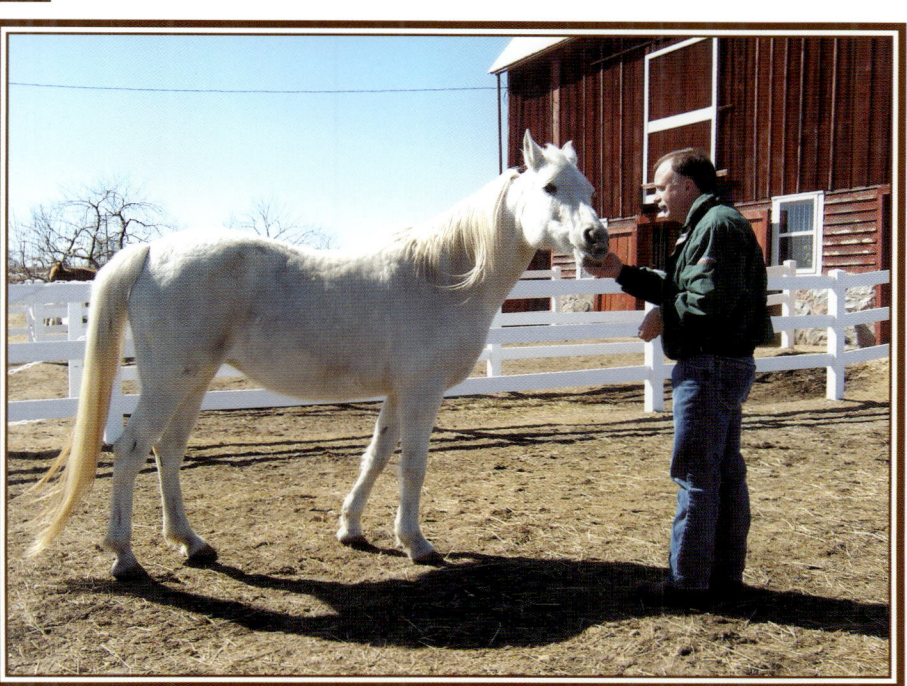

Information provided by Rick and Peg Frederick, Tim Carroll, and Whitney Lyon.

Mandy and Rick Frederick

Before restoration

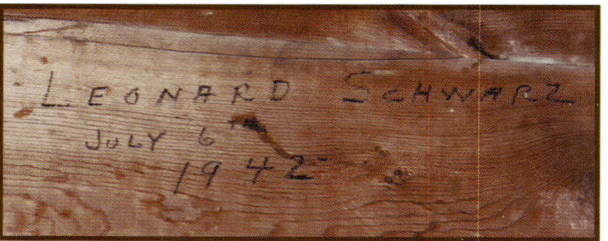

May Mills with her husband and baby in front of their house across the road

174

KRUPKA BARN

This little barn by the side of Smokey Hollow Road is a soft red with white flower boxes and cherries decorating the boxes. It has always been admired and photographed by passersby. But it is a barn, utilized for many things in its long life. It is an around-the-turn-of-the-twentieth-century barn, according to Ken Pickett, who has taken a great interest in it over the years.

In 1881, John and May (Golden) Mills acquired 40 acres with Smokey Hollow Road separating the property. The Mills house was across the road from the little barn. Today little is left to show where the house once stood. "May had written a saying over each door in the house. She was a very literary person and she exchanged books with my Grandmother Holmes," Becky Wells told me. "My grandma and she were great friends, I remember." May's husband died in 1929 leaving May and family to farm alone. She was said to be a strong and active woman in the community. She was still driving her car in 1960 when she was 92. She died later that year. In 1938, May had sold the property on the east side of the road, which included the barn, to Roy Hooper, who already owned other property next to May's.

Physical Description
Date Built:
- Unknown, but likely 1900 or so

Date Restored:
- ca. 1985

Construction:
- Plank frame
- Horizontal grooved siding
- Asphalt shingle gable roof
- Pier foundation made up of rocks

In 1985, Ken Pickett and his wife bought the house near May's former property on the west side of Smokey Hollow Road. Ken said his wife and he always enjoyed the abundance of wild flowers growing on this forgotten property. A large American chestnut tree still remains on the property with a metal plate attesting to the fact that it was a record breaker. Sadly, it is partially dead and is not going to live much longer. Ken admired the little barn by the side of the road, and when he heard that Terry Wells, the barn's owner at the time, was thinking of tearing it down, Ken asked if he could restore it. Restoring old barns was one of Ken's passions, having done this sort of work before. He put a new roof on the little barn, replaced rocks in the foundation to shore it up, painted it, repaired the windows, and put up the window boxes. He also planted daffodils and tulips around it. He cleaned out the inside, bringing the old building back to its original glory.

In the early years, the barn was used by the Mills family to house the horse and carriage. It was used for pickers quarters in the 1940's and 50's during cherry harvesting. There is a second level to the barn which could house 15 or so pickers. There is writing on the walls of the barn telling who had stayed there in the past. A small room on one side of the barn contains a cistern holding water run off from the roof. In that same corner, outside, is a well that provided drinking water and water for the horses. This was May Mills' only source of water for the house until at least 1948. One day, Bernie Rink and a friend saw her hauling water from the well by the barn while they were examining the large old chestnut tree in her front yard.

Terry Wells owned the barn for 40 years. In 1998, Jim and Fran Krupka bought thirty acres from the Wells family, including the little barn. They have added three more acres to their property. The little gable roof barn was repaired again when Jim shored up the foundation and straightened the building. He painted it a nice bright red too. The Krupkas built a new home into the hillside on high ground, and Jim, who already had two very important jobs, became a famer. He has planted eleven acres in grapes and harvested his first crop in 2005. While 10% were picked by hand, 90% were harvested with a mechanical picker. When asked what would be his title as a farmer, he declared himself to be a vintaculturist, in accordance with Michigan State University agricultural terms. His dream is to restore the land to an active farm, growing grapes and cherries. He also is a Deacon for St. Joseph Catholic Church on the Peninsula and managing director for the Chateau Chantel Winery.

Ken Pickett told me a good story about the year the National Governor's Conference was held at the Grand Traverse Resort in Acme in the mid-eighties. Everyone was sprucing up their buildings, as many of the nation's governors took tours of the Peninsula. He was on a ladder painting the little barn, and several cars stopped, the drivers wanting to know why he was painting it. Each time he would answer, "The governors are coming, the governors are coming!"

Information provided by Jim and Fran Krupka, Terry and Becky Wells, and Ken Pickett.

Before restoration

LIGON/TOMPKINS FARM

"Great-grandfather R.D. Tompkins, wife Sophia, and their large family—four sons along with several girls—came to the Old Mission Peninsula in the 1860's. They took up land from the government, planted trees and built very comfortable homes." Fida Tompkins Myers wrote this in a letter in 1994.

The Ligon Farm was known as the Tompkins Centennial Farm as it was continuously occupied by a Tompkins family member until 1979. The first Tompkins on the farm was Seth B. He came with four brothers from Titusville, Pennsylvania, down the St. Lawrence River, to the wilds of Northern Michigan.

Seth B. Tompkins worked for the Walker family on Old Mission Road when he first moved to Northern Michigan. He married Matilda and together they bought a farm on Old Mission Road. The 45 x 50 foot barn was built in the 1880's. It has a three wall shed that was added in the 1920's. The couple had three children, Murry, Rose, and Willy Gill. Rose and Willy Gill received land to farm while Murry farmed with his father. From 1880 to 1930 Seth raised the crops, fruit mainly, and Murry raised the livestock, cattle, and hogs. They also had a bull.

Physical Description
Date Built:
- 1880-3 wall extension added in 1920's

Date Restored:
- 1999

Construction:
- Timber frame with back extension
- Vertical wood siding
- Metal gable roof with cupalo on main barn
- Stone foundation

As I look at the 1908 plat map, I see different Tompkins names on many different pieces of property. After Seth B. died, Murry took over the farm and farmed with his son, Seth L. When Seth L. married Rebecca they purchased a piece of property from Guy Tompkins on Center Road. All that remains of this farm today, though, is a sturdy old stone wall.

The original Tompkins Farm was then farmed by Seth L.'s brother Doug and his father, Murry, who passed away in 1968. Murry's wife Lulu lived out her life on the farm. Seth L. managed the family farm on the original homestead until 1986, as well as his farm on Center Road. Seth and Rebecca's daughter Linda married Lenny Ligon, and they came to the family homestead in 1975. Lenny was a construction worker by trade and worked on repairing a small house on the farm for Linda and him to live in. He found he loved the land and soon gave up the construction trade for farming. He said he enjoyed the work and learning from his father-in-law, Seth. Seth had farmed all his life, and had a lot to offer Lenny in the ways of farming. In 1979, Lenny and Linda signed a contract to buy the property on a land contract, paying a $1000 down payment on the farm.

Now, Lenny and his second wife, Eddie, grow cherries, apples, and wine grapes on an 80 acre farm. They have a real mixture of animals on the farm…geese, chickens, turkeys, peacocks, 4 dogs, and horses. The Ligons have a pair of beautiful Percheron horses that they hire out for carriage rides. This farmyard reminds me of a small zoo. The buildings are delightfully painted with chickens on the chicken coop and horses on the barn doors. The whole farm yard is a very pleasant setting. There are two grapevine sculptures of rearing horses in front of the barn. A horse and wagon of grapevine and other "found" objects grace the front of the house. Eddie Ligon is an artist! A small chicken coop which Eddie's father made for her in the 40's holds special meaning for her. Painted with roosters, the coop is moved to wherever Eddie resides.

The barn was in need of repair, so a complete renovation was undertaken. The roof had a huge hole in it causing damage inside. Eddie said, "In 1999, we started from the foundation, which was washed out, and we worked up. It took one year to restore. Getting the junk out took the longest time!" The gable roof barn, complete with cupola and lightning rods, is beautifully restored. It has a metal roof and nicely painted vertical board and batten siding.

LIGON/TOMPKINS FARM

There is a large water tank near the barn, painted to look like it is made of large red bricks, with a Percheron horse painted on it. Lenny explained that every farm needed a water storage tank, because the well pumps of the day could not keep up with the demand for water for the sprayers. These were called rod pumps, and they needed to run continuously as it was essential that the tank be kept filled when so much water was needed. Lenny said that in the 1960's the orchards were sprayed "in dilute." This meant using 400 gallons of water per acre of apples and 300 gallons of water per acre for cherries. The orchards would need to be sprayed up to fifteen times a growing season.

Lenny remembers a story about his father-in-law, Seth L. It seems Seth got a new BB gun and had a really good time shooting out all the glass windows in the barn. When his dad asked him why he did it, he replied that he liked to hear the tinkle of the glass as it broke. It was not recorded what happened to little Seth after that.

Information provided by Rebecca Tompkins Nothstine and by Lenny and Eddie Ligon.

Chicken coop

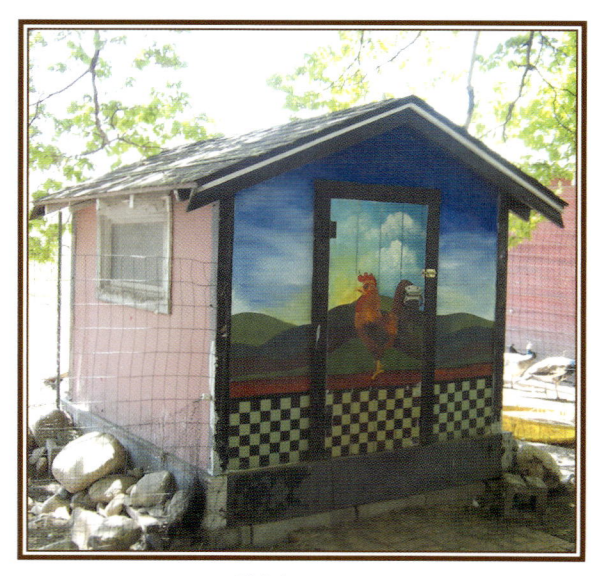

Chicken coop

Water storage tank

MINERVINI BARN

This barn was built around the turn of the twentieth century and has been completely renovated. Its several extensions make it a very long barn. Ray Minervini says it has over 8000 square feet. He bought it in 1994 and started redoing the old building, taking six months and about $50,000 to complete it. He said he worked from the bottom up, so the big red barn has a new foundation as well as a new roof. Inside, part of the space holds his spacious workshop. The rest of the barn is used for storage. His business is a builder and construction contractor. Ray's passion is renovating and restoring old historic buildings. He has worked on several in Traverse City, including Building 50 on the old state hospital grounds, and is doing a splendid job of bringing it back to a productive life.

But, let's start at the beginning. I talked to Reva Gore Newman about the farm. Both Reva and her father, Russell Gore, were born on this farm. Milton and Catherine Gore were the first that she can recall on the farm. Catherine was a teacher, and Milton, a man of many talents, came from Chicago where he was doing a stint in vaudeville. They met (Reva isn't sure where) and married, and moved to the Peninsula. Their first house was small and was just north of the present "new" house, which was built in 1911. Milton and Catherine had four children…Henry, Leslie, Russell, and Grace. Grace Gore Fouch lived to be over 101 years old, and passed away in 2006. Reva's father, Russell, told her that he and his two brothers slept upstairs in the "old" house. In the winter they would wake up in the morning to find on their quilts a thin layer of snow which had sifted through the roof during the night.

Physical Description
Date Built:
- 1900

Date Restored:
- 1994

Construction:
- Timber frame bank barn with glazed tile silo
- Horizontal wood siding
- Ribbed metal gambrel roof over asphalt shingles
- Concrete foundation

8,000 square foot barn - renovated

MINERVINI BARN

Before restoration

Reva said her father took classes in animal husbandry at Michigan State University and was always interested in horticulture. He added cement structures to the south and west of the barn. Here he made individual stalls for his cows, had a separate bull pen plus a hog pen. He had 10-12 Jersey cows, raised for the high butterfat content of their milk. The cream was sold for farm extras, especially in the Depression when money was tight. She and her mother Dorothy also made cottage cheese and butter.

The farm was 50 acres wth about 10 acres in mangolds (like sugar beets), hay and corn, which was used for cattle feed. The rest of the acreage was pretty much in cherries…both sour and sweet. They also grew apples. Reva said her dad liked to experiment with grafting to see what he could get, and she remembers he had one tree that had seven different varieties of apples on it. They also grew prunes, nectarines, peaches, plums, red raspberries, and boysenberries. In addition, the farm had Carpathian and English walnut trees as well as filberts and Chinese chestnuts.

At the back of the property in a wooded area, they had a maple sugar bush. When the syrup got to a certain stage in the boiling down, they would have a party and make taffy, cooling it in the spring snow around the bush. Reva remembers her father was pouring cement for an arch in the sugar bush house when they received the news that Japan had just attacked Pearl Harbor. Those of us old enough to remember know where WE were when that horrific news was broadcast on the radio.

Reva's job was to pitch silage down to the cattle inside the glazed tile silo that stills stands today at the east end of the barn. It has a wood sided top that adds to the uniqueness of the barn, as there are very few silos left on the Old Mission Peninsula. Reva said she got the job of pitching silage because her father was very allergic to the mold that would accumulate in the silo. Other jobs on the farm included pitching hay and driving tractor while Mom and Dad were spraying the orchards with hand held sprayers. The farm was a lot of work, with chores to be done twice a day. Through it all, Reva has many positive memories, like playing in the barn. The kids would climb high into the loft and swing from ropes. She said once when she went first, she did not let the rope get taut before she swung out, and it jerked her arms so hard she was sure they were pulled right out of their sockets. She said they all loved to play hide and seek in the barn too.

The family always had a roadside stand for selling food products and stuffed animals. Reva said they sold a lot of pies. They put tangle foot, which was sticky shrub, around the bottom of the stand to prevent ants from invading their food stuffs on the stand and had the curly, sticky fly paper hanging in the stand to keep the flies down.

When cherry picking season came, the family would get help from people from Flint who worked in the automobile factories who came up on vacation to help harvest the cherry crop. Also, "drifters" and Mexican families were hired. After 1941 the farm started getting more mechanized, and good crops of cherries started coming in.

Reva lived in the house until 1986, when she sold the land to Ward Johnson. This land was the first in the area to go into the PDR (Purchase of Developmental Rights). Ray Minervini bought the house, barn and three acres in 1994. When Ray and Marsha Minervini was renovating the barn, he found a whole series of cisterns around the barn. He said one cistern on the northwest side of the barn is so big, the kids could swim in it. This one was above ground, sloping west, and was filled from rain water run off from the roofs. Reva recalls, "My dad installed two sets of hand hold pipes around three sides so I could swim in it. Also well water was pumped by the windmill on the property in earlier days. We didn't have time to go to the beaches because of all the farm work in the summer time and there were always chores to be done. We were at least two and a half miles away from both of the public beaches with no way to get there."

Ray said that as he worked on restoring the barn back in 1994, farmers would drive by and express appreciation for saving the grand old barn. Ray told me, "From my perspective the barn has more charm, with the patina of age, that you just cannot replace with a new one."

Just down the road to the south was Reva Gore Newman's brother's farm. Leslie Gore raised cherries, and, of course, he had a barn. In 1963, when the tornado tore through the Peninsula, his barn was one that was hit. It was demolished, and today the elegant Chateau Chantel, an expansive French style winery and bed and breakfast stretches across the top the hill where the Gore farm once was. The beautiful villa is surrounded by vineyards. The entrance has an angel, made of grapevine, heralding its welcome to all visitors.

Information provided by Ray Minervini and Reva Gore Newman.

Before restoration

BRET and SONYA RICHARDS BARN

This farm was one of the Tompkins brothers' farms and was a showplace when it was owned by Willy Gill and Flora Tompkins. The barn was built in 1890 and the house in 1892. Bret and Sonya Richards have restored both of these old Peninsula buildings and now they are most pleasing to the eye.

This 36 x 42 foot gambrel roofed, plank frame barn housed cattle, sheep, and hogs at one time. The loft held hay and dried corn for cattle feed. There was no silo on this farm as there were not enough cattle to warrant it. Willy Gill and Flora farmed 60 acres. The Tompkins had two children, Grace and Oliver. Oliver acquired a farm south of his parents' farm. Daughter Grace married Van Gleason and helped run the Tompkins farm.

In the 1940's Van Gleason and Marshall Pratt, the neighbor across the road, ran a cherry weigh station on Pratt's farm. The cherries, after the weighing process, were taken to Ellsworth for processing. They also packed apples for shipping. All of the buildings are still in place. This farm is now owned by John Wunsch and Laura Wigfield.

Van and Grace Gleason had two children, William and Jean. When Van died in 1946, Willy Gill continued to live on the farm until his death in 1960. Grace Gleason married Waldo (Cub) Spruit in 1951 and went to live on the Spruit farm on Old Mission Road. Cub Spruit was well known in Traverse City as the football coach for Central High School. After Grace Gleason-Spruit died, son Bill Gleason and his wife Virginia came to farm the Tompkins/Gleason acreage. They had 40 acres of cherries and kept cattle, both milk cows and feeder cattle, in the barn until around 1982. They built a lovely home way back from the barn, almost unseen from Center Road. About 1986-87 when farming was no longer profitable, the land was split into four 12.5 acre plots and sold.

Before Bret and Sonya Richards bought their 12.5 acre parcel, the farm with the barn on it was rented out. Many boards from the side of the old barn were ripped off to keep the renters' stove going in the house in the winter. Bret has replaced many boards on the barn with the original hemlock siding. The siding is called board and batten because of the narrow strips of wood bridging the wide planks. Bret also had the barn reroofed in 1997 for $20,000. The barn has recently been painted and it is back to its "original magnificent self." This ramped barn has an unusually long bridge to the upper level. Today, the barn itself is used mainly for storage but stands out as a credit to the farming community as you drive along Center Road.

The farm is still a working farm, as the Richards own and operate Harbor View Nursery, growing annual and perennial plants, schrubs, and flowers in their greenhouse. They also have apple, cherry, and peach trees. Many of the small maple trees that you see as you drive along Center Road are from Harbor View Nursery. The Old Mission Historical Society under the leadership of Leo and Rebecca Tompkins-Nothstine have planted trees to help beautify the drive out the Peninsula. Since 1999 the Nothstines have planted 25 trees a year. Leo and Rebecca feel it is a valuable project and would like to see others get involved in planting trees along the roadside. Our lovely Peninsula was once totally forested. There still are 100-year-old maple trees along the roads, but they are dying off and becoming fragile, splitting and breaking in the wind. Thus, they have been cut down because of the potiental danger. Drive down Smokey Hollow Road and you will still see a tunnel of trees just before Ladd Road. Leo Nothstine feels strongly about the replanting project. He said, "If our forefathers thought enough to plant trees, often making shade for the horses as they trotted along the road, then we should think enough of the Peninsula to replace dead and dying trees."

Information provided by Bret & Sonya Richards and Leo & Rebecca Tompkins-Nothstine.

Physical Description

Date Built:
- 1890

Date Restored:
- 1996-97

Construction:
- Timber frame bank barn, ramped
- Vertical wood siding
- Metal gambrel roof
- Fieldstone and cement foundation

Before restoration

THOMPSON BARN

In the late 1800's this land was owned by the Traverse City Lumber Barons (probably Hannah and Lay) and consisted of 120 acres. The Lyon family, immigrants from Sweden, worked for the lumber barons, logging off the acreage. When the lumber was gone, the Lyon family bought the farm and grew corn, grain, and potatoes. The barn was built in 1905 from timber remaining on the land. It is a timber frame bank barn with a gambrel wood shingled roof. It still has lightning rods on the roof. One can only imagine the hard work that was required to clear the land of the stumps left behind by the huge trees before any crops could be planted. The sandy soil was not productive enough to make a good living. In the late 1930's the family gave up traditional farming and began growing cherries, like so many others on the Peninsula.

The original owner of the farm was Oscar Lyon. Next his son Charles, and later, Robert Lyon occupied the farm. Two houses of the same style were built side by side. Grandmother Lyon lived in one, and Charles Lyon with his wife and four children lived in the other house. Only one house remains on the property now. The other home was moved to the east end of Blue Water Road, and the Larimers now call it home.

The barn has been completely restored. It originally had 1 x 10 vertical board siding, but now is sided with 4 x 8 sheets painted gray. The barn has a fieldstone foundation and a drive through on the main floor. On one side it is ramped and the other side it is banked. The barn was used for cattle and horses on the lower level and hay storage in the loft. The barn also provided housing for cherry pickers at one time. There is a building close by that is called the garage, and it was used for fixing and storing machinery. Beverly Lyon, Robert's sister, remembers the farm well. She recalls that when the family had company, she and the other children got to sleep on the hay in the barn.

Physical Description
Date Built:
- 1905

Date Restored:
- 1992-93

Construction:
- Timber frame bank barn
- New vertical wood siding
- Wood shingled gambrel roof with lightning rods
- Fieldstone foundation

Recently the Thompson family has put 106 acres of their land into the PDR Program (Purchase of Developmental Rights).

Across Center Road from the Thompson and Carroll property is the Grand Traverse Chateau Winery, one of the first wineries on the Peninsula. From the scenic turn-off on top of Carroll Hill, you can see the sprawling buildings and vineyards which are part of the winery. In all seasons it is a grand site as you can see both bays and miles of vineyards, farms, and several barns.

Information from Mr. Thompson, Frances MaCaw, and Beverly Lyon.

January 18, 1906
Mr. and Mrs. Andrew Gilmore and six little children had a narrow escape from death last night when their residence at Mapleton was afire. Harvey Fowler was passing the residence and saw the blaze and immediately gave an alarm to the sleeping family. The children were carried from the house by neighbors. The house was entirely destroyed but most of the furniture was saved. The loss is about $700, fully covered by insurance.

Before restoration

VERBANIC BARN

What a change has taken place in this barn! The main barn building was built before 1870 and had several owners before the Verbanics bought the property from a Mr. Ribney in 1986. The original acreage was 160, but by the time the Verbanics bought the property, most of the land had been sold off and just 1½ acres remained.

The barn on the Verbanic property, now beautifully restored, is a most attractive addition to this Bed and Breakfast location looking out on West Bay. The barn was in need of much repair. The foundation was shored up, and the backside of the barn had to be resided from the foundation up about three feet. The main part of the barn has a gable roof, and there are three sided sheds on each end of the barn. At one time horses were kept in one of the sections. There are lovely gardens spread around the spacious yard. The barn adds immeasurably to this lovely setting on West Bay. It is used now for storage of garden equipment and as a big garage. There is an office and sitting area in one section.

Physical Description
Date Built:
- 1870

Date Restored:
- 1986

Construction:
- Frame with three sided sheds on each end
- Original horizontal narrow shiplap wood siding
- Asphalt shingle gable roof with cupola
- Foundation is shored up with cement and stone

One of the former owners sold land north of the Verbanics to the township to create Bowers Harbor Park. Voters had rejected the purchase of the park land in 1916, 153 to 53. Its destiny might have been much different had it not been for the foresight of Charles Lardie, a local farmer. He purchased the land with borrowed money. At the September 28, 1938 Township Board meeting, members voted to pay Charles Lardie $1,600 for the park, and $500 was given by the township board for improving it. Horse shows were held in the park in the past. There was a complete set-up arena and stands for the shows. Now it is a spacious family park with a field set up for ball games of all sorts, a playground area, a covered picnic area, and a walking, running, and hiking track around the perimeter of the park.

Next to the park on Bowers Harbor Road is the Bowers Harbor Winery with an elegant farm house and buildings for wine making and tasting beside it. The winery sits nestled among the rows of grapevines surrounding it.

Information and picture from Gary Verbanic. Other information from the Peninsula Township Board minutes and Laura Johnson.

Another view of the Verbanic barn

BAY BARN

This 1920 gable roof barn is 30 x 60 feet and has been redone with new siding since John and Ruth Bay bought the ten acres in 1986. The barn was covered with tarpaper with battens to hold it in place. Looking more like a large garage now, it is an attractive building. Ruth said, "We didn't know what color to paint it after siding it. Red didn't seem appropriate, so we painted it to match the house." It sits at the back of a lovely landscaped yard with statues and a gazebo on the side yard.

One of Joseph Kroupa's sons, Bernard, took over the farm in 1936, when his father bought the acreage. Bernard and Dorothy were married in 1937 and had three children who were raised on the farm. The barn held cattle with hay in the loft. It was also used to hold machinery. However, the barn was primarily used for cherry brining, one of the steps in the making of maraschino cherries. It is an on-grade barn and is used for storage now.

Growing up on the farm Bern Kroupa, son of Bernard, remembers there was no need for weight lifting classes in school when he was growing up, as lifting the hay bales into the loft of the barn gave him all the weight training he needed.

Information from Bern Kroupa and Ruth Bay.

Physical Description
Date Built:
- 1920

Date Restored:
- 1986

Construction:
- Plank frame with three sided addition
- Vertical wood siding
- Asphalt shingle gable roof
- Cement foundation

Before restoration

KROUPA BARNS ON KROUPA ROAD

KROUPA BARNS ON KROUPA ROAD

At one time there were seven Kroupa families living on Kroupa Road. The following stories and pictures of five barns tell about John Kroupa Jr. and Freda Nelson Kroupa who had four sons settle near their parents on Kroupa Road. The two barns and stories after that tell of John Sr. and two of his children, who also lived and farmed on Kroupa Road…

Information provided by David and Joan Kroupa, John Kroupa, Kathleen Kroupa, Deb Allen, Clarence Kroupa Jr., and from The Descendants of Simon Kroupa, *by Robert Ellis Schrader.*

First the barns and stories of the family of John Kroupa Jr.

David Kroupa's hand-made limb trimmer

Sadie

John Kroupa Jr. was born at Bowers Harbor in 1874, the oldest son of his parents John Sr. and Annie Zoulek Kroupa. He had this 30 x 40 foot barn built in 1910 with the lumber coming from the acreage, as did the rocks for the fieldstone foundation. It is a bank barn, has a gambrel roof and lightning rods of top.

DAVID KROUPA BARN

1910

Physical Description
David Kroupa Barn
Date Built:
- 1910

Construction:
- Timber frame bank barn
- Board and batten vertical wood siding
- Metal gambrel roof with lightning rods
- Fieldstone foundation

JOHN KROUPA III BARN

The beams in the barn are hand hewn and pegged. David Kroupa, grandson of John Jr., who lives on the farm now, said he was told when the barn was fairly new a storm came up and lightning struck. Grandpa John hightailed it for the house with the horse in tow, but the dog would not follow and was killed. Luckily just the barn's stone wall was cracked by the lightning, but there was no fire. The barn did not have lightning rods at the time.

John and Freda had nine children, four sons—Donald, George, Fred, and Clarence—all settled on farms nearby. They raised crops, had cattle and other livestock and were largely self-sufficient. Eventually cherry and apple orchards were planted and these became their main crop.

David and wife Joan live in the John Jr. family homestead. The barn is now used mainly for storage, picking ladders, cherry lugs, and the like. In the lower level a harvester is stored. Dave is very mechanical and has lots of machine parts around the yard that he will eventually recycle and turn into a new farm tool. I was told by Cal Jamieson that he most likely made the first one-man cherry shaker with some of his spare parts. Dave and Joan Kroupa participate in the Michigan Agricultural Environmental Assurance Program using improved environmental practices.

Physical Description
John Kroupa III Barn
Date Built:
- 1922

Construction:
- Framed ramped barn
- Horizontal wood siding
- Corrugated metal roof with lightning rods
- Fieldstone foundation

Dave and Joan's son John and family live on the farm nearby which belonged to Dave's parents, Donald and Evelyn Kroupa. Son John said he is the fifth generation Kroupa on a farm on Kroupa Road. John's barn is a 24 x 36 foot two-level metal gable roof barn that was built in 1922. It is mainly used for storage now, having once housed cattle and hay in the mow. His property also houses the winery where Peninsula Cellars wine is produced. The Kroupa family are owners and proprietors of the Peninsula Cellars Winery on Center Road in the old Maple Grove Schoolhouse. Dave estimates he has a little over 20 acres in grapes on this farm. Besides growing grapes, the family grows cherries and farms about 300 acres spread around the Peninsula.

This barn and several others are scattered along Kroupa Road. The barns were all built around 1910-20. Nearby stands the George Kroupa barn, covered with vines obscured by trees and brush and little used. It was used for cherry pickers' quarters before mechanical harvesters. There is writing on the walls from some of the former occupants of this on-grade barn with loft.

KATHLEEN KROUPA BARNS

Physical Description
Kathleen Kroupa's Two Barns
Date Built:
- Around 1910-20

Construction:
- Timber frame on-grade barns
- Vertical wood siding
- Metal gambrel roofs
- Concrete foundation

Just around the corner where Fred and Dorothy Kroupa once lived is another small gable roof, on-grade barn. It is not in use nor usable for much more than storing things. It does have an interesting star cut out at the peak. Both of these barns and farms are now owned by Kathleen Kroupa and the late Tom Kroupa. Together they comprise about 40 acres. At one time I am sure these barns housed cattle, horses, perhaps had a loft full of hay, but now have outlived their usefulness and are left to the elements.

"Spirit hole"

Another barn from John Jr.'s descendants on Kroupa Road is the David and Deb Allen barn. It was farmed by Clarence Kroupa Sr. and sons. Deb Allen believes the house and barn were built over a one-year period, around 1912. It is a wooden barn with galvanized steel siding in some places and a metal roof. It is 22 x 32 feet with a shed that is 16 x 32 feet. It looks like it has been well used in its time.

Clarence Kroupa Jr. said his father had 40 acres initially and added 40 more. He began to plant cherry trees in the open land at least around 1934. The farm was used for general farming, and Clarence remembers having a horse for cultivating corn and other crops. Two milk cows and feeder cattle were housed in the barn. There is a loft in the barn over the stalls and one on the other side which was a hay mow. Clarence did not like milking the cows and he said, "I never could feel comfortable with them and I was sure they knew it. I liked it when we could just buy milk!"

When he was 12 or 13 years old, Clarence did regular farm work, including driving the tractor and the spray rig in the orchards. At 14, he had a driver's license and hauled cherries into Traverse City to the canning factory. During difficult times (the Depression) he remembers the kids would pick up apples from the ground, or they would shake them from the trees with a pole that had a hook on it. These apples were taken to Morgan's Processing Plant in town to make cider.

Physical Description
David & Deb Allen Barn
Date Built:
- Around 1912

Construction:
- Timber frame on-grade barn with shed addition
- Vertical wood siding
- Metal gambrel roof
- Concrete foundation

Clarence left the farm when he was 25 years old. His brother Dewaine and family continued to farm. Clarence said he never lost his "farming" instinct and always has a big vegetable garden in the summer.

David and Deb Allen purchased the farm in 1996 and used the barn for storing a 1957 Chevy Bellaire. The barn is turned into a haunted house for the area kids at Halloween time.

DAVID & DEB ALLEN BARN

Now for the barns and stories of the two families of John Kroupa Sr. - Alfred Kniss and Carol Kroupa Zientik…

193

KNISS BARN

This nicely painted red barn was built in 1920. Alfred Kroupa, fourth son of John Kroupa Sr.'s second family, was born in 1899. He married Lois Tompkins in 1922 and they bought this 40 acre farm in 1936 from the Russell family. Thirty acres were on the barn side (south side) of Kroupa Road and 10 acres were across the road. Maxine Kroupa, daughter of Alfred and Lois, married Daryl Kniss in 1944. Daryl took care of the cattle on the farm when his in-laws were gone in the winter. In 1963, he took over the farm, but always lived just down the road. Daryl and Maxine's son Alfred and daughter-in-law Sue acquired the farm in 1986. Now Daryl and his second wife travel to the southwest in the winter. (His first wife passed away in 2001.)

The 40 x 50 foot barn has a gable roof with two levels and a poured concrete foundation. There is a shed attached to the east side of the barn with an open end once used for the white-faced Hereford to get "fresh air," as Daryl Kniss put it. He called it a manure shed. Because the farm could not grow enough hay for the cattle and had to purchase it, the Herefords were eventually sold.

Basically, it was a 30-acre cherry farm, also growing pears and peaches. During the picking season Daryl remembers there were 60 to 80 migrant pickers. Some of the migrants wanted privacy so they would pitch tents away from the farm in an open space. Most were housed in migrant buildings that were provided for them. Daryl remembers starting with a limb shaker in the 70's, but later acquired a trunk shaker from California. This shaker was really meant to shake pecan trees and was a bit too rough on the cherry trees. Cherries don't need a lot encouragement to drop from the tree. Daryl said they had to change the balance of the rotary device to make the weights lighter, thus toning down the force of the shaking. He told the salesman selling this shaker he was not buying it unless the heavy shaker could get up the hills that the orchards were on. The problem was solved when the machine was backed up the hill, as the back of the machine had more weight.

Physical Description
Date Built:
- 1920

Construction:
- Timber frame with large slope roof extension
- Vertical wood siding
- Asphalt gambrel roof on barn and addition with lightning rods
- Cement block foundation

Information provided by Daryl Kniss.

ZIENTEK BARN

This centennial farm was settled by John Kroupa Sr., one of the first Peninsula farmers, who homesteaded 80 acres on Kroupa Road. He was born in Bohemia in 1850, and came to American as a young child in 1854 with his parents, Leopold and Petronilla, who had come with four sons. Son John had two families by two wives. Annie Zoulek was his first wife who bore him nine children. John Jr. was the oldest child, after a sister died at 10 months of age. Four of John Jr.'s sons had farms along Kroupa Road. When Annie died young, John Sr. married Catherine Pasok in 1892 and together they produced nine more children. Perry Kroupa was the youngest of the second family. He always lived and worked his father's farm. His daughter Carol and her husband Robert Zientek bought the farm from him in 1974 and moved there in 1976. Perry Kroupa died the next year, and Carol said her husband lamented that his father-in-law did not have enough time to teach him all he needed to know about farming. Robert had a town job managing the Credit Union in Traverse City. Clarence Kroupa Jr. told me that Perry was a very good farmer and made the most of the land. Woods and a small lake take up part of the property. As time went on much of the acreage was put in cherries. Carol said her father never took to working with mechanical harvesters, the cherry shakers, and was unhappy that technology had taken over.

This 28 x 52 foot gable roof barn has a metal roof and was built in 1910. Carol remembers her father raising hay and corn for the few cattle kept in the lower level. She said the cattle were allowed to roam around in the barn in a cattle pen in the winter. The barn still has hay mows on both sides of the upper level. As with so many other farms in the fifties, the farm was almost totally self-sufficient. Carol remembers that as soon as she and her sisters got home from school, her father could be heard telling them, "You sat all day, now get your work clothes on and get busy." Now the 76 acres of cherries are farmed for Carol by another farmer. The barn is used for storage but once housed cherry pickers and cherry supplies.

Physical Description
Date Built:
- 1910

Construction:
- Timber frame bank barn
- Vertical wood siding
- Metal gable roof
- Fieldstone and cement foundation

Carol's husband Robert kept every building in good working order and it is a beautiful homestead. He died suddenly at the age of 62, and Carol now lives on the farm alone, though her son Robert lives nearby. She told me the lovely farmhouse was once a duplex, as her father and his wife lived on one side, and Grandmother Kroupa lived on the other.

A good story about Carol's grandfather John Sr. involves how he found his second wife, Catherine Pasok. When his wife Annie died at 38 leaving him with nine small children to care for, John took a trip to Detroit looking for a mother for his children. He was a very handsome man and he soon met a young Bohemian woman who was scrubbing floors in a bar. He asked her to marry him on the spot. She needed a bit of time to think, but they married in 1892. Carol isn't certain if Grandpa John told her about the nine children he had at home, but nevertheless he whisked her away to the north country. She went on to have nine children of her own, kept house, and did all that was involved with living on a farm in the 1890's-1900's. Carol says this wife was a tiny woman though of sturdy stock and strong conviction.

Oral information from Carol Kroupa Zientek and from The Descendants of Simon Kroupa by Robert Ellis Schrader.

OLD MISSION PENINSULA FARMING & COMMUNITY THROUGH THE YEARS

Early Grain Binder
Photo courtesy of Jack Holman

When the first settlers came to the Old Mission Peninsula in the mid 1850's, they found the local Indians had learned to grow wheat and potatoes from the government specialists who were sent with Reverend Dougherty (ca.1842). The newcomers soon learned that the soil and the climate were ideal for growing potatoes and apples. Not only could the land produce these products in abundance, but the unique location of the Peninsula, with its convenient access to Lake Michigan to ship their produce easily, helped develop a thriving agricultural community. Potatoes were widely grown on the Peninsula until around 1910-20. By not taking care to rotate crops, though, the potatoes wore out the fragile sandy-loam soil.

Orchards and vineyards were to replace the potatoes. Due to the unique climate conditions found on the Peninsula…warm water on two sides, gentle breezes caused by the thermals on the hilly terrain, and the warm sandy loam…the Peninsula enjoyed a growing season approximately 30 days longer than lands farther inland on the northern part of the lower peninsula, and there was less risk of frost.

Before the turn of the twentieth century, the *Chicago Herald* printed an enthusiastic article (October 1892) about the Old Mission Peninsula, "garden spot of Northern Michigan." In part, it read, "More fruit was shipped to Chicago than any other port in Northern Michigan. Barrels of apples, baskets of peaches and pears, boxes of plums, grapes, etc. One thousand bushels of raspberries, 100 bushels of cherries, and 500 bushels of winter apples particularly have obtained the highest price in the Chicago market. Two thousands bushels of potatoes excellent in size and quality were sent by steamer last year also. The fishing business is nearly overshadowed by farmers and resorters."

Apples were grown in abundance and shipped in barrels from Old Mission Harbor or Bowers Harbor from 1880 well into the 1900's. These were Duchess, Northern Spies, Wagners, Russets, Jonathans, Delicious, and Greenings. In 1889 H.K. Brinkman developed an evaporation plant on his property in Old Mission in which apples were dried. Also, about this time Charles Reese ran a cider mill at the corner of Swaney and Mission Road. (Information taken from *The Insider's Guide to Michigan's Traverse Bay/History*.)

In 1905, cherries became increasingly the fruit of choice and local farmers continued to expand their cherry orchards until, at one time, Old Mission Peninsula had the greatest concentration of cherry trees in Northwestern Michigan. The tart cherries were Montmorencys and the sweet cherries grown were Napoleons, Schmidts, and Windsors. Cherry trees are planted 100-108 to an acre, do not bear fruit until the fourth year, and do not reach profitable production until the seventh year. An average crop of cherries per acre yields approximately 10 tons of mature tart cherries. (Cherry facts from a brochure compiled by the Bowers Harbor Junior Women's Club) A need arose for weighing and processing the cherries on the Peninsula. A story about these businesses is also included in this chapter.

Other fruits grown on the Peninsula are pears, prunes, and now an ever increasing number of grapes for wine. In 1974 Ed O'Keefe planted the first 35 acres of vines, growing grapes specifically for wine making. He took quite a risk, but now there are five successful wineries on the Peninsula. Mr. O'Keefe's winery is called Grand Traverse Chateau. By 1993 another winery, Chateau Chantal, opened and sold its first bottle of wine. It now has 98 acres of active vineyards. Once again the combination of natural and social factors make this area an ideal region to grow fine wine grapes. One acre of grapes produces between 3-4 tons of grapes, and 3 tons of grapes yield about 500 gallons of wine. It takes six years from the initial vine planting to reach commercial production of the wine. (Information from Jim Krupka, manager/director of Chateau Chantel) Other wineries found on the Peninsula are Bowers Harbor Winery, Peninsula Cellars, and the newest, Brys Estate Vineyards and Winery.

Today, we have a tendency to consider over half of the Peninsula as an ever increasingly scenic, comfortable, and lovely place to live, with an occasional barn having farm land up on the higher (middle) ground. However, when you reach Mapleton, about 11 miles out, the scene begins to change to the way things were…farms, older homes, and more history. Here is where Old Mission Peninsula becomes a tourist destination as thousands of visitors come to see the Old Mission Lighthouse and the historic Hessler log cabin at the tip of the Peninsula. The tiny hamlet of Old Mission still has the original general store. About 400 yards north of the store is the Mission House built in 1842 by Reverend Peter Dougherty, the first "stick built" house between the Straits of Mackinaw and Muskegon. It is a real treasure, and the local historical society and other interested persons are working diligently to see that it is secured and renovated so that all may enjoy yet another piece of Old Mission history.

Harvesting Potatoes—In the early 1900's the Peninsula was potato country. Every year the school was dismissed for two weeks so the children could be home and help dig potatoes, called a potato digging vacation.
Picture courtesy of Grand Traverse Pioneer and Historical Society

Farm life—1900's
Picture courtesy of Grand Traverse Pioneer and Historical Society

The land after it was logged off.
Used for pasture or crops when stumps were pulled.
Picture courtesy of Grand Traverse Pioneer and
Historical Society

Old Mission
Sorting apples for cider or evaporation plant

E.O. Ladd apple orchards
Apples are packed into barrels—1909

Cherry picking on Boling Farm
Photo courtesy of the Brauer family

Guy Tompkins Cherry Orchard. Clipped cherries were boxed and ready to go to Traverse City for shipment to Chicago by rail. They were iced to keep them cold.

A young cherry orchard with intercrops. Other crops, like corn, were planted between rows of small cherry trees, utilizing the soil until the trees grew up.

Picture courtesy of Grand Traverse Pioneer and Historical Society

Cherry harvesting sequence…
from the tree to the receiving station and cooling pad.

The cherry pitter was invented in 1917 by Mr. Dunkley and is still the same today.

Pictures courtesy of Heather Reamer

1.

2.

3.

4.

5.

6.

201

PENINSULA FRUIT RECEIVING AND PROCESSING PLANTS

Axel Ostlund and a group of growers started the Peninsula Fruit Exchange. While there are other receiving and weigh stations on the Peninsula, the Peninsula Fruit Exchange is the only fruit processing operation. It was organized as a stock company in 1952. Jim Horton, the current manager, said he started working at the exchange in 1965 when he was 16 years old. He said cherries were still being brought in on the trucks in lugs. Axel was the manager of the company, as was Don Shea, one of the other growers in the company. Jim points out that they always seem to slide from one job to another without a real defining job as manager.

Jim Horton pointed out that there is a reason for the multi-tasking, as the Fruit Exchange is a very busy place in the summer, usually right after the Fourth of July. For the next four to five weeks the trucks come rolling in filled with cherries in 1000 pound tubs. They are unloaded and then go on to various plants to be canned or frozen. Graceland Fruit and the Smeltzer Company are companies that process dried tart cherries. Other fruits—apples, plums, and pears, when in season—are sent to Gerber Baby Food Company for processing into applesauce, strained plums, and pears.

From October to April about 60% of the cherries are processed for the making of maraschino cherries, Jim said. One thousand pound bins of sweet cherries are dumped into wooden tanks where they are bleached and firmed up with preservatives. The stems are removed, the cherries are graded by size, and the pits are removed. The bleached, pitted, sized cherries then are sent on to finishers, at another plant, to be dyed green or red. The largest ones are packed in jars. Smaller ones may be used for chocolate covered cherries, and those that are no longer whole, but still good, are put into ice cream. The Fruit Exchange has about 25 workers at the present time.

Lorey Kroupa weigh station

There have been a number of cherry weigh stations on the Peninsula through the years. Gleason and Company, which was founded in 1946 as Pratt and Gleason, was a receiving station for fruit. Kroupa's Inc. was founded in 1947, and started the brining process for the maraschino cherry business as well as processing tart cherries. R.C. Warren and Company was also a brining operation on the Peninsula Fruit Exchange property from 1952 until 1985 when the Exchange purchased the operation. Today only the Johnsons have a receiving station on Center Road.

Between 1920 and 1960 Cherry Growers, Inc. had a receiving station on Peninsula Drive and Bowers Harbor Road and was organized to handle the large amount of cherries needing to be processed from the Peninsula. The weigh station's platform was built to the height of a semi-truck flat bed allowing fruit growers to bring in their load, have their cherry lugs removed, get their ship weigh slip, and be on their way. The truck from Cherry Growers in Traverse City then would back up to the platform and carry the load into town. It is now a historic building that is no longer needed. The scale, which weighed the lugs of cherries, cannot handle the size of the cherry tanks used after the advent of the cherry shaker. Lorey Kroupa, proprietor and owner of the historic weigh station, adapted the use of the station to a fruit, vegetable, and flower market. In 1982, with the assistance of Governor Bill Milliken, Lorey had the weigh station and the attached four acres placed on Michigan's Historic Register.

Peninsula Fruit Exchange

1. Planting vines

2. Ripened grapes on the vine

3. Mechanical harvester

4. Mechanical harvester

5. Mechanical harvester

6. Harvested grapes dumped into tubs going to winery

Pictures courtesy of
Jim Krupka

A DAY IN THE LIFE OF A FARM WIFE
BY MARY JOHNSON

Up before it's light outside!

Fix a HUGE breakfast for the family—ham, potatoes, applesauce, hot biscuits, and anything else left in the cupboard.

Wash all the dishes by hand. (Thank goodness for dishwashers now.)

Help with milking the cows—squirting a little bit into the cats who are sitting there watching every move you make and hoping you haven't forgotten them.

Strain the milk; save the cream in the five gallon milk can because the dairy pays a little bit for it, which helps. Feed the strained milk back to the chickens and pigs. I've tried making cottage cheese, but it never tasted like Walter's mother's, so the animals enjoyed it all.

Get the children off to school.

Time to wash all those little discs that strained the milk. If there is enough cream, now make butter…on an old fashioned butter churn. Straighten up the house and make the beds.

Start weeding the garden, and perhaps find enough vegetables for dinner and supper. Dig around the base of the potato plants for the little potatoes—they are delicious at this size. Green beans are prolific, too, and they always taste good from the garden. Maybe there is a squash, too.

If apples are ripe make a couple of pies, or take some cherries out of the freezer for a treat. (My mother-in-law showed me how to bake pies. Hers were so-o-o-o good.)

If you baked bread yesterday, just add some excellent homemade jam for a taste treat. Mother Johnson also had bees and would collect honey. That was really good with hot bread.

Dinner and supper were the same big meals, just change the meat to beef or pork, add fresh vegetable and those pies you made earlier, and everyone had a big meal.

Bring the cows back across Center Road from the meadow behind the house to the barn to be milked.

Sometimes I drove tractor to help Walter pick up the lugs of cherries from the Mexicans after they picked the trees—before shakers.

By dark, we were all ready for bed. It had been a typical day for us.

In spite of the work, I loved being here, and doing all the things I'd never done before. I never heard my husband complain about anything, either. Maybe he was surprised that a city girl could adapt so well to the country life.

Mary Johnson married Walter Johnson in 1946 and lived and worked on an 80-acre farm, raising four children. The children, two girls and two boys, live on the Peninsula and the sons, Dean and Ward, have followed in their father's footsteps, as farmers and fruit growers. Both girls are writers.

THE FIRST FIRE TRUCK

Rather than let a house burn down that was just across the road in Peninsula Township, the Traverse City Fire Department extended its services beyond the city limits in the 1920's. The city began charging for these calls around 1930 and the Township paid for them. It seemed unfair to residents on the north end of the Peninsula to pay for services not available to them, due to the distance, rough roads, etc. The Township Board voted to discontinue paying for the fire calls in 1932.

Through the efforts and contributions of the Cherry Center Grange, later called the Township Hall, a homemade fire truck was assembled in 1943. Sponsored entirely by the Grange, a volunteer fire department was formed with Stanley Wood as fire chief. The truck was built from parts of a spray rig, a truck chassis, a tank, and a Bean pump (an industrial strength pump made by John L. Bean Co.). It had ladders attached to the sides. John Lardie and Ben Hager put this apparatus together at the Mapleton Garage, and the Peninsula Fire Department was officially started.

The main problem with this first fire truck was that it was not large enough. Also, when the tank was filled with water, the truck would not go very fast but it worked well for a while and put out quite a number of fires. (From *Reflections of Yesterday* by Julianne Meyer)

In 1944, the Township Board agreed to provide a garage, if the Grange finished paying for the truck. Voters approved the purchase of a new fire truck in 1947 and an official township volunteer fire department was organized.

Currently the Fire Department consists of two stations, one at Mapleton and one near the base of the Peninsula. The number and type of vehicles have changed considerably. Rich Vandermay listed them as two tankers, two engines, two grass fire trucks, two ambulances, one water supply truck, and one rescue truck. With Rich's able volunteer crew, our Peninsula is well protected.

Harry Heller, Arnold White, Roy Hooper, Claude Watson, Ike & John Lardie, Stan Wood, Ray Heller

GROWING UP ON THE PENINSULA
BY GEORGE MCMANUS JR., RETIRED STATE SENATOR

I was born on the Murray Farm in the 1930's. In 1935 the folks bought Uncle Alfie McManus's farm just north of the Frederick Tower on Carroll Hill. So, we were natives of the Archie community, which is that area of the Peninsula south of the Swede Crossroad. (Island View Road now.)

After we moved onto the farm we had electricity installed because there was a line down Center Road. There were no lines on the shore roads. At this time Center Road was gravel. East Shore and West Shore were not much more than two tracks. Then, a little later we had a wall phone from the Mapleton Phone Company installed. It was a party line affair with a side crank to get the operator if you wanted long distance. That meant any place in the world south of Carroll Hill! Our number was 14F2. When we answered the two rings usually several others also came on the line to "catch the news." One long ring indicated a fire so everyone took the message, dropped everything to go and help put out the blaze.

Primary fuel for heating and cooking was wood. One daily job was to make sure kindling to start the fire and wood to burn were in the woodshed. Water for cooking, cleaning etc. came from either rainwater collected in a barrel from the eaves on the house or from the well, which was located halfway between the house and the barn. About 1940 we piped water to the barn, and later piped it into the house. My grandparents had no well, only a spring.

Originally, there was no bathroom in the house. An outside two holer was standard fare with thunder mugs for winter use. Bathing was done with a wash basin during the week, and on Saturday nights a thorough washing was done in the wash tub in the kitchen. A full bathroom was not installed until the end of World War II…about 1945.

There was no TV. We had a radio so Dad could listen to the fights, and later during the war, he listened to Gabriel Heater at supper time, with the 6:00 news. The Lone Ranger, The Green Hornet, and The Shadow were serials we sometimes listened to on rainy days. Traverse City did not have a station until 1939 when Les Biederman brought in WTCM. Before that we only had WJR Detroit and WGN Chicago. On a clear Saturday night we would listen to the Renfro Valley Boys do mountain music from Tennessee.

When we were old enough, about nine, our job was to milk the cows, A.M. and P.M. Then we would separate the milk, feed the skim milk to the pigs, feed the horses and the chickens. We did all that before and after school and all the other days too. Summer jobs included cultivating corn, hauling hay, picking cherries, as well as hoeing in the potatoes and the garden.

Our school, Maple Grove, was one mile south of our farm. It was a Kindergarten through seventh grade one-room classroom. Everyone walked to school….winter, spring, and fall. Classes began at 9:00 A.M. Recess was 10:30; lunch was at noon. Recess again at 2:30, and school was out at 4:00. Our recess time was 15 minutes. Noon lunch was an hour. Mrs. Lenora Pitcher was our teacher. She made $50 a month. Most teachers were getting $35 a month, but the school board, which included my grandfather, Arthur McManus, felt she deserved $50. And, she did not have to do the janitor work which some teachers also had to do, for another $15 a month.

There were 35 students at school. The county Superintendent, George Eikey, showed up once a year. Miss Farr, the health nurse came once a month. We went through the usual childhood illnesses…mumps, measles, chicken pox, and of course, head lice. Head lice were cured by washing your hair in kerosene at home, and then picking out the nits by hand. At one point we had a vaccination day when the nurse set up an "office" in the basement. The school children and their family members were invited to come for shots. A pair of kerosene lamps were set on the table. They had open flames…no globes. The needles used for the vaccine were passed through the open flame to sterilize them between each shot, being re-used for the next patient!

"Outside" speakers were invited to make presentations at the P.T.A. meetings. One I especially recall was a Finnish lady, Lydia Kotalonin, who gave a talk on her visit to Lapland. She brought pictures and artifacts from that far off land. Every school had a 4H program. Our leader was Ed Bopry, who taught us elementary carpentry. Ed also had a boat with a light, used for spearing suckers in East Bay in the spring after dark. If you were in 4H, you got to go spearing with Ed and sometimes ride in the boat. Suckers were plentiful, and we ate them smoked, fried, or pickled. My Grandmother McManus also steamed them over boiling potatoes.

Our neighborhood was heavily Swedish and Irish with a few other nationalities. Most of the Irish were relatives, and many second generation Peninsula families intermarried. It was important to know who was related to whom. I was surrounded by aunts, uncles, great aunts and uncles, grandparents, and cousins. All of them were helpful on occasion, and, kept my parents informed.

Sunday was a day off. Catholics attended Mass at St. Joseph Church near Mapleton. The services were 8:30 A.M. the first and third Sunday, and 10:30 A.M. the second, fourth, and fifth Sunday of each month. The priest had Elk Rapids and Barker Creek churches also. Mapleton and Barker Creek were missions of Elk Rapids, as in the early days the priest could travel by water or on the ice. Methodists attended the Ogdensburg Church. The Congregationalists were at Old Mission. The Swede Church had closed before my time, so they attended services in Traverse City.

Winter recreation for many of the older folks in our neighborhood included card playing. My grandparents were experts and taught me Michigan Rum, 500, and Euchre. They rarely could afford a new deck of cards so as the cards got more used and often sticky, we applied talcum powder to keep them slippery. My folks and couples their age carried on Saturday night card parties, rotating from house to house, with the hostess putting out a midnight lunch. This was done only in winter of course. There was no time for cards in the summer.

No one had much money in the 1930's. Because everyone was poor, it didn't bother us that much. Health suffered somewhat. As there was no health insurance, people avoided hospitals, doctors, and dentists as much as possible. Cherries were the "cash crop" but they were only bringing a cent and a half a pound, so the cream check from the cows was important.

Migrant workers who came to help harvest the crop were from other parts of Michigan and eventually Kentucky, Tennessee, and Arkansas. Texans came later. For us, a vacation consisted of a trip to the Upper Peninsula after cherry season to "see the sights."

In the 1940's economics began to improve. Cherries went to 4 cents a pound, the OPA ceiling price, so more amenities came along. After the war, the ceiling price was lifted. In 1945 and 1946 cherries went up to the astronomical price of 15 cents!! This changed the Peninsula forever. Diversified farming with livestock went by the wayside and agriculture became specialized. Roads were paved, utilities modernized, and life would be defined as "much better." More and more students began finishing high school in Traverse City, with several (including me) going on to college.

However, the lessons learned in the one-room school, the oral history repeated by grandparents and others who had time to "sit a spell" and talk, and the 1930's economic lessons plus living in a now-recognized beautiful area makes me thankful for those early beginnings.

Maple Grove School has become Peninsula Cellars Winery

ESSAYS FROM FOURTH GRADE STUDENTS AT OLD MISSION SCHOOL

I

The Peninsula is pretty different because water surrounds it on three sides. East and West bay are on the left and right of the Peninsula. Many people enjoy living on the Peninsula with all of the barns, plant and wild life making the Peninsula special in Michigan.

Why should we preserve the barns of the peninsula? Barns are a window into our past. People make beautiful art with just looking at the barns and drawing what they see. They are good for storing horses and hay. If there is a storm then the horses might get wet and the hay could get blown around. These are some reasons why we should preserve barns on the peninsula.

Should we preserve plants? If people keep plucking flowers from the ground, well, some times they don't grow back. The peninsula should preserve the plant. I want to keep the peninsula beautiful so people can see how pretty it is when they come up.

Does the peninsula need to preserve wildlife? Animals such as birds get shot by people a lot. Many people just shoot birds for fun and they do not use the bird they just waste it. So birds should not get shot by people. Every bird deserves to fly free and get throught life. I hope this message changes everybody.

What I like about the peninsula is it's beauty. By Samantha Knudsen

II

We should keep the Peninsula preserved. The Peninsula is located in Grand Traverse county, up in northwestern Michigan. It is a beautiful piece of land and I don't want houses, apartments, and factories to take it over.

One reason we should keep the Peninsula the way it is is the barns. The barns have been around a long time. Some of them were here before Traverse City! Now that's a long time ago! They are a window into the past! I bet the people in two generations will ask "How old is the peninsula?" And, they will learn how old the barns are and they will probably say, "Wow, the Peninsula is old. I'm glad people preserved the barns" But, if you put houses, factories and apartments on it, then the people might not ever know how old barns were.

One core Democratic value for this situation is "Common good!" Not making the peninsula more populated is good for the people already living here. So that is why we shouldn't put other houses, factories, and apartments on the peninsula. By Chris Cooke

These essays were created for the Old Mission Writing Program under the direction of Karen McClatchey, 4th grade teacher. Mrs. McClatchey taught at Old Mission School from 1990-2005.

OLD MISSION PENINSULA
LEADER IN FARMLAND PRESERVATION

Old Mission Peninsula is one of the most unique agricultural areas in the world. Surrounded by the west and east arms of Lake Michigan's Grand Traverse Bay, the Peninsula's rolling hills are ideally suited for growing a variety of fruits including, and especially cherries, and wine grapes. Ironically, these same hills and ridges with their spectacular views of the Bay are also a magnet for development.

Given this situation, Old Mission Peninsula was at a crossroads in the late 1980's: would this magnificent landscape suffer the fate of so many other areas and be overwhelmed by residential subdivisions? Or, alternatively, would its citizens take action to save what is most precious and dear, thereby channeling development to more suitable locations?

For those of us who cherish our rural way of life, the good news is that over the past fifteen years or so, Old Mission Peninsula has become a national leader in farmland preservation efforts. This in part is because of the work of three private land conservation organizations: American Farmland Trust (AFT) along with the Old Mission and Grand Traverse Regional Land Conservancies. The other key ingredient to the farsighted farmland preservation efforts on Old Mission Peninsula has been the visionary planning efforts of Peninsula Township backed by residents who have passed not one but two successful purchase of development rights (PDR) millage votes in 1994 and 2002.

All told, some 9,500 acres, more than half the total acreage on the Peninsula, have been targeted for protection through being designated part of an Agricultural Preservation Zone. Since the first PDR vote was passed, the inevitable tide of development has been stemmed, as only two farms on the entire Peninsula have been subdivided while some 3,765 acres of land have been forever preserved by this unique private/public partnership of land trusts working hand-in-hand with the township.

Some of the most notable success stories include establishing the Old Mission State Park at the tip of the Peninsula when the former Murray Farm was threatened by a massive subdivision development and the Gateway to the Peninsula scenic view protection project near its base along Center Road.

The success of farmland preservation efforts on the Old Mission Peninsula has resulted in a powerful statement that not only does agriculturel matter, it is here to stay. Moreover, there has been an unanticipated side benefit. With the rise in confidence that agriculture is here for good, many of us have begun to make infrastructure investments, including preserving the scenic barns that help to define our cherished rural character on the Peninsula. Farmland protection and barn preservation efforts are proceeding hand-in-hand, which is why I am so pleased with Evelyn Johnson's efforts to document the history of our barns on Old Mission Peninsula and the special beauty of these magnificent structures built with love and care by pioneering families. How fortunate we are that both the land and these beautiful barns will be preserved for the enjoyment of generations to come.

Glen Chown, Executive Director
Grand Traverse Regional Land Conservancy

BIBLIOGRAPHY

Avery, Julie, Stephen Stier, and Jack Worthington, *Barns of Michigan's Thumb*, from Michigan Folk Life Annual. MSU Museum. 1998.

Beckett, George, Swaney Family History. 1982.

Cracker, Ruth, *The First Protestant Mission in the Grand Traverse Region*. Rivercrest House. Mt. Pleasant, MI. 1979.

Leach, Dr. M.L. *A History of Grand Traverse Region*, written for the *Grand Traverse Herald*. Traverse City, MI. 1883.

Lyon, Mary R. *St. Joseph Catholic Church History 1850-1996*.

Johnson, Walter and Julianne Meyer, *Memories Hidden, Memories Found On the Old Mission Peninsula*. 1993.

McManus, George Jr., Memories of Living On the Old Mission Peninsula Archie Community.

Meyer, Julianne, *Reflections of Yesterday*. 1988.

Mohr, Nancy L., *The Barn Classic Barn of North America*. Courage Books. Running Press. 2001.

Murray, David, Editor, *Greetings From Peninsula Homefolks World War II - 1944-46*. Compiled by Jack E. Solumnson.

Noble, Allen G. and Hubert G.H Wilhelm, *Barns of the Midwest*. Ohio University Press. Athens, Ohio. 1995

Potter, Elizabeth V., *The Story of Old Mission*. Edwards Brothers Inc. Ann Arbor. 1956.

Schrader, Robert Ellis - Compiled - *The Descendents of Simon Kroupa*. 1976.

Sprague, Elvin L. Esq., and Mrs. George N. Smith, *Sprague's History Grand Traverse and Leelanau Counties*. Unigraphic Inc. Evansville, Ind. 1903.

Stadtford, Curtis K., *The Land and Back*. Charles Scribner & Sons. N.Y. 1972.

Stier, Stephen, *Michigan Barn Farmstead Survey Manual*. MSU Board of Trustees. 2000.

Wait, S.E., *Old Settlers of The Grand Traverse Region—1918*. Reprinted 1978.

Wakefield, Larry, *Butchers Dozen—13 Famous Murders*. A & M Publ. Co. W. Bloomfield, MI. 1991.

West, Bonnie, THE STRIP. 1974.

Westfall, Joanne M. *Architecture and Site Design Guidelines*. MSU Press. 1997.

Wilber, Addison, *Memories*. 2001.

Wilson, Robert, *Grand Traverse Legends—Biographies of The Early Years. 1838-1860. Volume I*. Grand Traverse Pioneer and Historical Society. 2004.

A Talk With Katherine Bagley Marshall—Interview by Reverend James Brammer. September 6, 1976. Transcribed and designed by Laura Lee Marshall. Dec. 1990.

With These Hands—A Celebration of Farm and Family In The Heartland. Grand Traverse Regional Land Conservancy. 2000.

With permission, quotes from *The Record Eagle* and *Grand Traverse Herald, 100 Years Ago*.

Index, with Photographs in **Boldface**

Allen, David & Deb Barn 193; Deb, 190
Altenburg, George 141 & Helen 10, 141
Alvarez, Anne 75
American Farmland Trust 135, 209
Anam Cara Farm 169-171
Andrus Family 59, 99; Jon 45, 99
Archer, Joseph 22, 45
Archie Community 206
Archie Community Women's Club 3
Archie Hall 57
Atherton, Mary 47
Auger, Edgar, Grace, Jerry 18
Ayers
    Gladys 5; Willis 8
    Ted 5-6, 79, 95
Bacon, Ed & Grace 10
Bagley
    Emma Pratt & Harriet Parmalee 59
    Ted & Lucile 79; William 59
Bagley Pond 58-59
Baird Barn 2; Brett 3
Baldwin, Luther & Nellie 13
Barns, Types of 120
Bay Barn, 188; John & Ruth
Beckett, George 5, 95
Bee, Barbara (Andrus) 5, 146, 148; Tom 6, 95
Bee, Debra/Russ Patton Barn 94-5
Bee, Dennis & Karen, Barn 4-6
Beers Barn 92-3; Mack & Lorraine 93
Benson
    B.A. 43, 121
    O.J. 43, 75, 121, 123, 153, 173
Benson, O.J., Barn 121
Boch, Jane Griffin 8-9
Bohemian Cemetery 111
Boling,
    George, Sr. & Ethel Hill 127
    George, Jr.& Frankie, Robert 127
Bonney, Joan 121
Bopry
    Bob & Barb 117; Ed 61, 117, 206
Bos Family 57
Bos/Montague Barn 56-57
Boursaw,
    Bruce 71; Jane Louise 45; Tug 109
Bower's Harbor Inn 99
Bower's Harbor Park 69
Braddock, Jesse 84-85
Brauer Barn 126-127; Rich & Marty 127
Brill, Dorothy 69
Brinkman, H.K. & Oliver 21
Bryant, Jim & Fern 99
Brys Farm 97; Walter & Eileen
Buchan
    Elizabeth (Carroll) 125
    Frank, Lester, William E., William H. 123
    Norm & Karen 123, 168
Buchan Farm 23, 122-33, 168

Buell, Robert & Martha Belle 22, 45
Burkhardt, Frank 83
Carlson, Jon 99
Carmody, Patrick 127
Carroll,
    Al 90, 138; Alex & Daisy (Johnson) 125
    Andy, Albert, Helen 173
    Bernie 28; Charles 28, 145
    Edward & Jane Holman 125, 133, 173
    Fred 125, 131; Lawrence 84, 90; Ray 84
    Stephen, Mary, & Celia 84-5
    Timothy 125, 173; William 29
Carroll Barn 124-5
Cauchy, Charles 8-9
Cauchy/Griffin Barn 9
Cherry Center Grange 205
Chown Barn 10-11; Glen 209 & Rebecca
Christopher,
    Frank 33; Harry 65; Perry & Dale 67
Church Crossing Road 125
Cilke, Ruth Nelson 93
Citizens Telephone Co. 35
Colburn, Eliza Elizabeth 21
Cooke, Chris 208
Cooledge, Charles, Fred, Ken, Lillian 90
Cosgrove Family 43; James 143
Cosgrove/Kitchen Barn 14
Cotner Barn 12; Vic & Ronni 13
Cowen, Mae 135
Crampton, Elmer & Gertrude, Wanda 151; Norm 45
Crampton Farm 150-51
Crandall, Eva, Deronda "Chum," & Max 31
Cross, Jim & Chris 93
Cutler, Barn 101; Richard, Carol, Katie
Dafoe Barn 58-59; Darrell, 59
Dayton Barn 60; Jon & Amy Jo
Devol, Edgar 55; Robert 6,13,21,55
Dohm
    Elton, Faye 27
    Fred, Sr. 14-16, 27, 67, 109
    Frederic, Jr. 14, 16
    Ray 131; William 14, 67
Dohm, Faye/Elton Barn 26-27
Dohm, Frederic & Kay Barn (Twin Maples) 14-16
Dougherty, Peter 141, 197
Dunne, Debbie Christopher 65
Edgecomb, Frank & Stella 155
Edmondson
    Dave 28-29, 47, 118
    Harold & Elsie (Sundeen) 118
Edmondson Barn 28
Edmondson Farm 121
Eldred, Richard & Marilyn 66
Elliot, Holly Haven 93
Elliott, Jeanette (Shambaugh) 10
Elzer
    Arnold 39, 75, 97
    Betty, John & Nancy 39

Elzer Barn 39
Erickson Farm 102
Far Out Farms 98-99
Feiger Barn 102; Walter & Susan
Field Barn 64-65; Dennis & Sue 65
Flachsman Barn 17; Sherry & Woody
Florence Barn 30; Ben & Janet 31
Fouch,
    Clifford 105 & Grace (Gore) 105, 179
    Dan 105, 109
    Howard & Ann 105; Perry 137
Fouch, Dan Barn 104-05
Fowler
    Chris & Colleen 107; **Joy 51**
    Curtis & Louisa 132-33; Zeb 134
Fowler Farm 106-07
Fowler/Holman Barns *132*-33
Franklin, Sol 99
Frederick, Rick & Peg 173
Frederick Barn 172-173
Frederick Tower 28, 29
Friday, Victor 51
Fulmer, Dr. & Karl 103
Germaine Barn 103; Charles
Giles Farm 97
Gilmore, Miles 14; Thomas 8
Gleason & Co. 61, 202
Gleason Family 89; Bill & Ginny 183
Goff Barn 32-33; Lyle & Evelyn 33
Goodman, Georgia 3
Gore
    Henry, Milton & Catherine 179
    Leland 151
    Leslie; Russell & Dorothy 179, 181
Gray
    Alan 167
    Albert P. 49, 167 & Elizabeth 167
    Edwin P. 167; Ellen (Sunquist) 121
    Floyd 103 & Lewis 103, 167; William 66
Gray & Co. 40
Gray/Springer Barn 166-67
Griffin
    Dudley 8-9; John, Leah, Florence 9
    Jayne 8; Jim 105
Grothe, Robin 81
Hagelstein Family 99
Hager, Ben 205
Halpin, Chris 66
Halstead, Lena Griffin 105
Harmon Barn 62-63; Bill & Judy (Herkner) 63
Helfferich Family 105
    Archie 67, 109; George 71, 109
    John & Charlotte, John Jr. 109
Heller, Lucie, Barn 108-109; Ray 205
Hemming,
    Jed, John, & Virginia 129
    John Jr. & John Sr. 35
Hemming, John Jr. Barn 34
Hemming, John Sr. Barn 128-29, 137
Herr Barn 18
Herkner, Ozzie & Etta 63

Highsaw, Jane 77
Hoffman
    Irene 61, 130
    Tom 61, 75, 90, *130*-31, 133, 173
    William, John & Matilda (Lardie) 130-31
Hoffman, Bill & Monica Barn 61
Hoffman, Tom & Irene, Barn 89, 130-31
Hollister, Donna 9
Holman
    Jack & Georgia, Bernard & Ruth 133
    Tim & Laurie, Cory 134
Holmes,
    Jack & Carol 13, 38
    Jean 87; John 13, 37, 133
    Katherine 13; Lula 37, 87
    Roy & Wilimina 37; Russell 13, 38
Hooper
    Roy, Jr. 98; Roy 87, 175, **205** & Theda 87
Horton Barn 135; Jim, Jr., Jim, Sr., & Arlene
Hyslop, Martie 74, 99
Jackson Barn 40; Ronald
Jamieson
    Cal 19, 71, 81, 117, 137; Verla 137
    George, Floyd & Mrytle 71; Bud 81
Jamieson Barns 136-37
Jensen, Angie 35
Jerue, Blossom Seaberg 7, 145
Johnson
    Ace (Captain) 99; Frederick (Captain) 129
    Dean 19, 69, 155-157, 204
    James M. Dr. 7; Richard & Mary Kaziah 125
    Laura 69, 187; Lester 155; Paul 29; Teri 19
    Mary 28, 157, **204**
    Walter 155, 162, 204; Ward 181, 204
Johnson, Dean & Laura Barn 68-69
Johnson, Dr. James M., Barn 7
Johnson, Walter & Mary, Barn 154-157
Johnson/Johnson Barn 19
Kauer; Ann (Zoulek) 17; Louis 17, 143
Keenan, Deirdre & Mary 6
Keith, Michelle 171
Kelley, George 61; Paula 21
Kelley Barn 20
Kitchen
    Leona 19, 143; Roger & Joe 143
Kniss Barn 194; Daryl & Maxine (Kroupa); Al & Sue
Knudsen, Samantha 208
Konig, Ben 106-107
Kroupa
    Adolph "Ade" & Thelma 110-111
    Albert 159; Bill 111
    Bern 38, 188; Bernard & Dorothy 188
    Bert, (Frisky & Harbor) 111, 159
    Bud **158-160**; Naomi 159-60
    Charles, Sr. 16, 111; Clarence, Jr. 190
    Dave & Joan 191; Donald & Evelyn 191
    Elizabeth & Henry 75; John III 191
    John, Jr. & Freda (Nelson) 190
    John, Sr. 111, 190, 195 & Annie (Zoulek) 190
    Joseph 75, 131, 188; Julius 111
    Kathleen 190, 192; Perry 195; Tom 192

Kroupa, Adolph Barn 110-11
Kroupa, Bud Barn 158-160
Kroupa, Dave, Barn 190
Kroupa, John, Barn 191
Kroupa, Lorey, Weigh Station 202
Krupka Barn 174-5; Jim & Fran 175
Ladd
    Elisha P. 36, 89
    Emmor O. Ladd 36, 83, **199**; Mrs. E.O. 13
Land Grant Certificate 12
Lardie Family 67, 103
    Anna Belle 79; Harriet 6
    Charles 113, 163, 187; Pete 143
    Elizabeth 162; George 13; Ike 35, **205**
    Jill, Judy, Ken, Chuck, Mary 163
    John **205**; Oakley 131, 162-3;
    Oliver 19, 10, 162
Lardie Barn 162-63
Lardie Store 8, 13, 77
Larimer, Russ & Deb 113, 185
Larimer Barn 112-113
Lee, Dr. Bill & Jeri 31
    Libby & Tom 3
Lehto Barn 70-71; Carl & Suzanne 71
Leighton Family 63
Leonard, Frank, Jr. & Sr. 47
Levin, Molly 10-11, 140-42
Lewis, Carolyn Johnson 155-157
Ligon, Lenny & Eddie 177-78; Linda 177
Ligon/Tompkins Farm 176-78
Lobdell, Greg 99
Lundeen, Evelyn 99
Lynch, Bill 31
Lyon
    Beverly 185; John 84, 153
    Alfred, Mary, Frank, Jack 153
    Oscar, Robert 185; Charles 113, 185
    Whitney 83, 121, 153, 173
Lyon Barn (Island View Orchards) 152-53
MacAlary, Viola 71
McCaw, Frances Carroll 28-29, 185
McKinley, Captain James 22
McManus
    Addie 90; Alfie 206
    Art "Big" 31; Arthur 31, 206
    George, Jr. 31, 83, 85, 90, 113, 133, 148, 165,
    206-07; Letita 31; Susie 41; Tom 17, 41
McManus Barn 41
McMullen Family, 49; John 51
Manigold
    Ken, Jr. & Sr. 115; Rob 65
Manigold Barn 114-115
Mannor/Holmes/Ladd Barn 36-38
Maple Grove School 191, 206-**207**
Maple Syrup 168
Mapleton Phone Company 206
Marsh, Jamie 98-99, **Barn 100**
Marshall
    Della, John, Jules 10
    Katherine Bagley 59; William A. 89

Mathison
    Ed 60; Frank 60, 106-107; Lavonne 107
Maveedy, Frank 99
McClatchey, Karen 208
Miller, Lewis 13
Miller Barn 42-43; William & Shirley (Cosgrove)
Mills, John & May (Golden) 175
Minervini Barn 179-81; Ray & Marsha
Minnema, John 83
Mission Hill Barn 44-45
Montague Family Barn 56; Richard, 57
Mohrhardt Family 117
Munson, Bob & George 3
Murray
    Ben 146, 148;
    David, Sr. 146, 148 & Carol 106
    David II 146, **148; Jane 148**
Meyer, Julianne 162, 205
Myers, Fida Tompkins 177
Myers, Roger & Martha 90
Neah-ta-wanta (definition) 160
Neah-ta-wanta Inn 47
Nelson
    Albin, Philip, John, Jennie (Swanson) 93
    Ludwig 93, 153; Oscar 69
Newman, Reva Gore 179
Nothstine, Leo 183
Novak, Claudia 45
O'Brien Barn 67; Barry & Laura
Ocanas Farms 138-39; Leo & Carmen, Emily
Ogdensburg Church 19, 206
Ogdensburg School 71, 151
Old Mission Dock 59, 141
Old Mission General Store 22
Old Mission Inn 35
Olsen, Charlie 81
Olsen, James & Karen 99
Olsen/Jamieson/Tiefenback Barn 81
Ostland, Axel 202
Oxteby Family 21
Panter Barn 72; Bob & Jo Ann
Parmalee, George 59, 146
Parmalee Cemetary 58
Patton, Russell 95
Pelizzari Barn 73; A.I., Eugene, John
Peninsula Fire Department 205
Peninsula Fruit Exchange 31, **202**
Peninsula Telephone Co. 35
Pickett, Ken 175
Pitcher, Lenora 206
Porter, A.E. 83
Porter House 35
Potatoes 198
Pratt
    Carl 10; Jerome & Araminta 141,
    Marshall, 183 & Mrs Marshall Pratt 89
    Will & Mary 141; W.R. 83
Pratt/Altenburg/Levin Barn 140-42
Pratt-Gleason 89, 183, 202
Prescott, Dr. 74

Prussing, Walt 63
Quinn, Tim 171
Rabine, Mary Ellen 81
Reamer, Heather Johnson 19
Reese, Thomas T. & Alice (Archer) 22, 45
Reinhardt, Jeff 75
Rheimheimer, Dean & Alexandra 3
Richards, Bret & Sonya, Barn 182-183
Richards, Jim & Marcy, Barn 22
Ridgewood Barn 147
Ridgewood Farm 146-48
Riley, Joe, Mary, Ann 165
Riley/Wilson Barn 164-65
Rose Farm 46-47
Rose Ridge Barn 54
Rosi Farm 76-77; Bob & Penny
Rude, Walter & Ruth 69
Ryckman Barn 74
Sabot, Sue 3
St. Joseph Catholic Church 121, 125, 162, 207
St. Joseph Religious Center 90
Santucci, Louis 75; Marc & Deb 61
Santucci/Kroupa Barn 75
Samuelson, Clarence 72
Schaeffer, Bellmont 81
Schelde Enterprises 99
Schlipt, Logan 99
Seaberg
    Anna, Jenny, Opal 7
    Blossom (Dillon, Jerue) 145
    Bob 61, 111, 117; Erick 7, 145
    Gust 144-5; Monica 61
Shambaugh, Dr. George E. 77
Shantz, Richard & Margaret 3
Shavey, Pat & Mabel (Helfferich) 109
Shea
    Don & Millie (Youker) 3, 31, 49
    Michael 49
Shea Barn 48-49
Shultz Barn 23; Gerald & Mary
Siefkin Family 77
Snyder Barn 116-117; Gene & Jean 117
Sobkowski, Steve & Nikki 141-42
Solomonson, Jack, 6, 38
Sondee, Mary Lynn 127
Springer, Jerry 167 & Barbara (Gray) 103, 121, 167
Sproule, T.N. & Elmira 81
Spruit, Waldo "Cub" 171, 183 & Grace Gleason 183
Steavens, Frank & Mary 151
Stier, Steve 120
Stickney Family 99
Stone
    F.W. 13; Lizzie 36-7
    Wilmina 37; William 13, 36-7
Stoney Beach School 31, 49, 167
Sundeen/Edmondson Barn 118; Ole Sundeen 118
Swaffer Barn 66; Fred
Swaney
    Charles, Frances "Fannie," & Rob 95
    Elmer, Harriet, Leslie, Gerald & Ann 79
    George, Jack, James, Lewis 5, 79
    John & Rosannah 5, 95
    Manasses & Mary Ellen 95

Swaney Barn 78-79
Swaney Lake 5, 141
Swartz, Dr. F.G. 84
Swede Crossroad 206
Swedish Church 121, 207
Taft, David & Sara 47
Teahen Barn 50-51; Jim, Roberta, Liz, Rebecca
Thiel, Alice Buell 22
Thompson Barn 184-85
Tiefenbach Family 81
Tompkins
    Doug, R.D., Sophia, Matilda 177
    Guy 8, 177, **200**; Murry & Lulu 47, 177
    Seth B. 47, 177 & Seth L. 37, 177-78
    Willy-Gill 177, 183 & Flora 183
Tompkins-Nothstine, Rebecca 13, 37-38, 47, 177-78, 183
Uithol, Gill 165; Nancy 163
Umlor, Ernie 19; Eugene 162
Urtel Barn 82-83; Jill 83
Van Cleve, Wallace, Jr. & Wallace III 31
Vaught, L.O. 77
Verbanic Barn 186-7; Gary 187
Vida, Rick & Rosie 3
Vida House 2
Vogel, Helen 53, 165
Wait, Enoch & Jennie, Eugene 98-99; S.E. 21
Walker Family (Sam) 115, 177
    Mary M. 77; Eloise & Susie 115
Walt's Barn Antiques 102
Warren, Elmer 66, 161; Ebb 161
Warren, Gary & Wendy, Barn 84-85
Warren, Keith & Jean, Barn 161
Watson, Claude 205
Welhusen, William 22, 45
Wells
    Dana, Floyd, & Helen 24
    Michael & Betsy 87
    Terry 87, 175 & Becky 37-8, 87, 175
Wells, Dana, Barn 24
Wells Family Barn 86-87
Westphal, Dr. Joanne 54
Wheelock
    Art. 148; Howard 75
White, Arnold 205
Wigfield, Laura 89
Wilber, Addison & Floyd 77
Willobee
    Abel & Florence 87; Menton, 71
Wilson
    Arthur, William 53
    Louise, Helen, & Beatrice 47
    Peter, James, Willard 53, 165
    Willis E. & Rose (Tompkins) 47
Wilson/Vogel Barn 52-53
Wineries, 197
Wolf, Pat 79
Wood, Stanley, Jr. & Sr. 205
Wunsch, John/Laura Wigfield Barn 88-89
Zientak Barn 195; Carol (Kroupa) & Robert
Zoulek; Tony 39; Peter 17
Zupin, Joan Crandall 31